AN INTRODUCTION TO ALFVEN WAVES

The Adam Hilger Series on Plasma Physics

Series Editor: Professor E W Laing, University of Glasgow

Forthcoming titles

Radiofrequency Heating of Plasmas
R A Cairns

Transition Radiation and Transition Scattering
V L Ginzburg and V N Tsytovich

The Adam Hilger Series on Plasma Physics

AN INTRODUCTION TO ALFVEN WAVES

Rodney Cross

Plasma Physics Department,
University of Sydney

Adam Hilger, Bristol and Philadelphia

British Library Cataloguing in Publication Data

Cross, R. C. (Rodney C)
An introduction to Alfven waves
1. Plasmas. Electromagnetic fields.
Oscillations
I. Title
530.4′4

ISBN 0-85274-245-2

Library of Congress Cataloging-in-Publication Data are available

Published under the Adam Hilger imprint by IOP Publishing Ltd
Techno House, Redcliffe Way, Bristol BS1 6NX, England
242 Cherry Street, Philadelphia, PA 19106, USA

Printed in Great Britain by J W Arrowsmith Ltd, Bristol

CONTENTS

PREFACE

It is now almost fifty years since Hannes Alfven first postulated the existence of Alfven waves. During that time, a vast amount of work on the subject has been published in the laboratory, solid state and space physics literature. This book does not pretend to cover all this material or even to review all the highlights. Rather, I have tried to provide an introduction to the subject at a level suitable for young research students. For these students, the existing literature is somewhat overwhelming in its volume and complexity. The main purpose of this book, therefore, is to provide a relatively simple introduction for beginners. Much of the material is in fact quite recent and may contain some surprises even for experts in the field.

The reader wishing to pursue the subject further will find plenty of material in the references at the end of the book. Nevertheless, even the elementary properties of Alfven waves can lead to heated debate among experts. The shear or torsional wave is especially troublesome. It is often stated, for example, that this wave is transverse magnetic, left-hand polarised, coupled to the compressional wave in an inhomogeneous plasma, refracted by a density gradient, has a continuous spectrum and does not satisfy a dispersion relation so it cannot exist as a wave. These statements are sometimes correct, given the right set of conditions, but it is equally possible to find conditions where none of these statements is correct.

This book has been strongly influenced by the observed behaviour of Alfven waves in the plasma physics laboratory at Sydney University. The laboratory was set up in 1961 by Charles Watson-Munro and Max Brennan. One of the main areas of interest at that time was the study of Alfven waves. That interest continues to this day. I thank these colleagues for introducing me to the subject and also my colleagues Neil Cramer and Ian Donnelly for many fruitful discussions over the years. I also acknowledge the contributions made to my understanding of Alfven waves by research students at Sydney University, especially Frank Paoloni, Boyd Blackwell, George Collins, Gerard Borg, Tony Murphy, Louis Giannone and Mark Ballico.

Since this book is intended mainly for research students, it has been prepared from camera-ready copy to minimise printing costs. I thank Miriam Pollock for typing the original manuscript and teaching me how to edit it. Jim Revill, Commissioning Editor for Adam Hilger, helped considerably by responding rapidly to my queries with many FAX and electronic mail messages. My wife, Voula, and daughter, Elizabeth, have been exceptionally patient with this venture, and to them I extend a special thanks.

Rodney Cross

∇ OPERATIONS IN CYLINDRICAL COORDINATES

$$\nabla p = \frac{\partial p}{\partial r}\,\hat{\mathbf{r}} + \frac{1}{r}\frac{\partial p}{\partial \theta}\,\hat{\theta} + \frac{\partial p}{\partial z}\,\hat{\mathbf{z}}$$

$$\nabla \cdot \mathbf{b} = \frac{1}{r}\frac{\partial (r b_r)}{\partial r} + \frac{1}{r}\frac{\partial b_\theta}{\partial \theta} + \frac{\partial b_z}{\partial z}$$

$$\nabla \times \mathbf{b} = \left(\frac{1}{r}\frac{\partial b_z}{\partial \theta} - \frac{\partial b_\theta}{\partial z}\right)\hat{\mathbf{r}} + \left(\frac{\partial b_r}{\partial z} - \frac{\partial b_z}{\partial r}\right)\hat{\theta} + \frac{1}{r}\left(\frac{\partial (r b_\theta)}{\partial r} - \frac{\partial b_r}{\partial \theta}\right)\hat{\mathbf{z}}$$

PHYSICAL CONSTANTS

Speed of light	c	2.9979×10^{8} m/s
Electron charge	e	1.6022×10^{-19} C
Electron mass	m_e	9.1095×10^{-31} kg
Proton mass	m_p	1.6726×10^{-27} kg
Boltzmann's constant	k	1.3807×10^{-23} J/K
Planck's constant	h	6.6262×10^{-34} Js
Permittivity of free space	ε_0	8.8542×10^{-12} F/m
Permeability of free space	μ_0	$4\pi \times 10^{-7}$ H/m

FOR A HYDROGEN PLASMA

$$v_A = \frac{B}{\sqrt{\mu_0 n m_i}} = 6.90 \times 10^{6}\ \text{m/s} \quad \text{at } B = 1\text{T},\ n = 1 \times 10^{19}\ \text{m}^{-3}$$

$$f_{ci} = \frac{Be}{2\pi m_i} = 15.25\ \text{MHz} \quad \text{at } B = 1\text{T}$$

1. INTRODUCTION

Alfven waves were first predicted to exist by Hannes Alfven in 1942 (B1) and were subsequently observed in laboratory plasmas in the late 1950's (C1,C5,C14). Not surprisingly, our knowledge of Alfven waves has expanded enormously in the intervening 30 year period. Most laboratory experiments with Alfven waves are now concerned with the physics and engineering issues of heating plasmas to conditions of thermonuclear interest. The most spectacular results have been obtained in large tokamak devices where ICRF (ion cyclotron range of frequencies) heating has been used to generate temperatures exceeding 5keV (~50 million Kelvin). Nevertheless, plasmas are sufficiently complex that there are still many issues waiting to be resolved and effects waiting to be discovered.

The main purpose of this book is to provide an introduction to Alfven wave propagation, at an *introductory* level. No attempt has been made to review wave heating results or to describe the more advanced theoretical aspects of the various heating schemes. Excellent reviews of this work can be found in the references at the end of the book (eg A5,F4,I5,I12,J2). Nevertheless, the primary motivation of most plasma physicists studying Alfven waves is concerned with plasma heating. A brief overview of the main heating schemes involving Alfven waves is therefore given in Section 1.2.

Another major area of Alfven wave research is in space physics, as described in Section 1.3. Despite the gigantic differences in length and time scales between space and laboratory plasmas, the physics of Alfven wave propagation remains the same. The only significant difference, in many cases, is the fact that laboratory plasmas are surrounded by metal walls. There are of course other differences, such as the typical values of magnetic field strength and plasma density, but most of the material in this book is just as relevant to space plasmas as it is to laboratory plasmas.

1.1 Physical properties of Alfven waves

There are two distinct Alfven wave types, known as the torsional and compressional Alfven waves. The torsional wave is also known as the shear or slow Alfven wave or simply as "the Alfven wave". At frequencies near the ion cyclotron frequency it is usually called the ion cyclotron wave. When kinetic effects are important it is called the kinetic Alfven wave. The torsional wave does not propagate at frequencies above the ion cyclotron frequency, at least in a "cold" plasma. In a "warm" plasma, the ion cyclotron wave *does* propagate at frequencies above the ion cyclotron frequency, as described in Section 12.7.

The compressional wave is often referred to as the fast wave. At frequencies well above the ion cyclotron frequency, the compressional wave is more commonly known as a helicon or whistler wave, while at frequencies near the electron cyclotron frequency it is called the electron cyclotron wave. The names torsional and compressional provide a clear description of the fluid motion of the two wave types. These names are generally used in this book in preference to "shear" and "fast". The word "shear" is appropriate in Cartesian coordinates, while "torsional" is more appropriate in cylindrical geometry.

The physical properties of the two wave types, at low wave frequencies, can be understood by noting that the $\mathbf{j} \times \mathbf{B}$ force in a plasma is equivalent to an isotropic pressure $p = B^2/2\mu_0$, together with tension $T = B^2/\mu_0$ per m^2 along the field lines (A2, p66). The torsional wave is closely analogous to a wave on a stretched string and therefore propagates along field lines with a velocity

$$v_A = \sqrt{T/\rho} = \frac{B}{(\mu_0\rho)^{1/2}}$$

where $\rho = n_i m_i$ is the ion mass density. More precisely, $\rho = n_e m_e + n_i m_i$, where $n_e m_e$ is the electron density, but it is essentially the <u>ion</u> density which determines the Alfven speed since $m_e \ll m_i$. The most distinctive feature of the torsional wave, consistent with the stretched string analogy, is that wave energy is guided along magnetic field lines. Consequently, if the wave is generated by a small antenna immersed in the plasma, the wave will propagate as a ray only along field lines passing through or near the antenna. If the wave is excited by an extended source and if ρ varies with position across field lines, as it usually does, then the wave front can change its shape with time since each section of the wave front travels along **B** at a different speed (the *local* Alfven speed). The latter effect is known as phase-mixing and is examined in Chapter 14.

The compressional wave involves compression of plasma transverse to **B** but not parallel to **B**. The steady magnetic field lines are also compressed (the torsional wave simply bends the field lines, without any compression). There is no plasma motion parallel to **B** since the component of the $\mathbf{j} \times \mathbf{B}$ force parallel to **B** is zero. Furthermore, in a collisionless plasma, there is no coupling of fluid motion in directions parallel or perpendicular to **B**. Since the compression is two dimensional, the phase velocity is given by $v^2 = \gamma p/\rho$ with $\gamma = 1 + 2/f = 2$ for $f = 2$ degrees of freedom. Hence $v^2 = B^2/\mu_o\rho = v_A{}^2$. If the plasma kinetic pressure is included, then the phase velocity of the compressional wave in a direction perpendicular to **B** is given by $v^2 = a^2 + v_A{}^2$ where a is the acoustic speed. The compressional wave is usually called the magnetosonic or magnetoacoustic wave when the kinetic pressure is considered to be significant. In this book it is generally assumed that $v_A \gg a$, in which case the compressional wave, at low wave frequencies, is isotropic and propagates with speed v_A at any angle with respect to **B**. Consequently, wave fronts from a point source immersed in an infinite, homogeneous plasma, will be spherical. Conditions where $v_A < a$ are commonly encountered in space plasmas, but not in laboratory plasmas. The propagation of Alfven waves under conditions where $v_A < a$ is considered specifically in Section 12.7.

In a laboratory plasma, the compressional wave is reflected at the plasma boundary or from the walls surrounding the plasma and therefore propagates as a waveguide mode or eigenmode. In a cylindrical or toroidal cavity, the wave can form cavity modes provided the wave frequency is above the waveguide cutoff frequency and provided that wave damping by the plasma is relatively weak. In some experiments, especially those concerning wave heating in plasmas with two ion species, the wave may be so strongly damped that cavity modes are not observed. If the plasma is separated from the walls by a vacuum or by a low density region, compressional waves can propagate as surface wave modes. These modes do not experience waveguide cutoff and therefore propagate at frequencies down to zero. Surface wave modes are described in Chapters 8 - 11.

Unlike the torsional wave, the compressional wave is refracted by density gradients. The wave energy is focussed towards the centre of the plasma since the wave is refracted towards regions of high density (see Chapters 7 and 14).

The fact that the torsional wave does not propagate across steady magnetic field lines has important implications for wave heating experiments. If an antenna is located at the edge of a plasma then it may generate torsional waves which propagate along field lines in the edge plasma, thereby heating the edge plasma. It is therefore important, both in the Alfven wave and ICRF heating schemes, to ensure that most of the energy coupled to the plasma propagates away from the antenna as a compressional Alfven wave (or a Bernstein wave). For example, at frequencies much less than the ion cyclotron frequency, antenna current elements perpendicular to the steady field do not couple to the torsional wave (G1). Since the torsional wave is transverse magnetic (at least at low wave frequencies), it is excited much more efficiently by current elements parallel to **B**.

1.2 Wave heating schemes

Most laboratory wave heating experiments are now conducted in tokamak devices. The basic features of a tokamak are illustrated in Fig. 1.1. The plasma geometry is toroidal (doughnut-shaped) with major radius, R_o, and minor radius, a. The largest device constructed to date is the JET tokamak with parameters $R_o = 3$ m and $a \sim 1.2$ m. A strong toroidal field, $B_\phi = 3.4$ T in JET, is produced by a set of external coils surrounding the vacuum chamber. The toroidal field is not uniform over the plasma cross-section, but varies as 1/R where R is the major radius coordinate. This result follows directly from Ampere's law, $\oint \mathbf{B} \cdot \mathbf{d}\ell = \mu_o I$, where I is the current in the external coils. The plasma is heated by a current, I_p, in the toroidal direction, up to 7000 kA in amplitude and up to 20 seconds in duration in JET. This current is induced by changing the flux through an iron core passing through the central hole in the torus. In other words, the plasma acts as a single turn secondary winding around an iron core transformer.

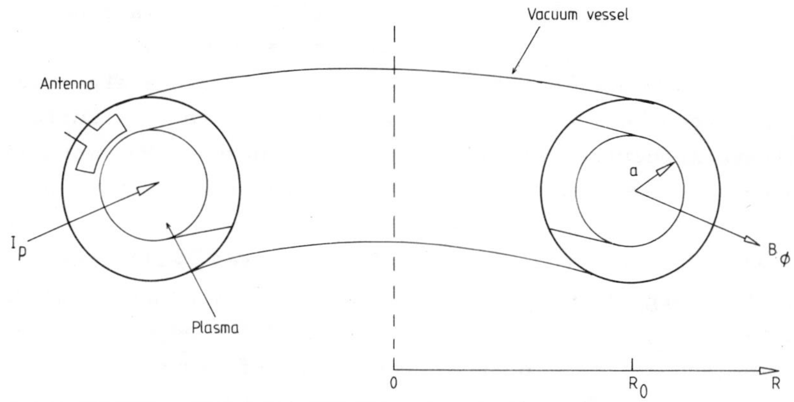

Fig. 1.1 Tokamak geometry. R_o = major radius, a = minor radius, B_ϕ = toroidal field, I_p = plasma current. Wave launching antennas must be mounted near the vessel wall, in a region of low plasma density and temperature.

The poloidal field, B_θ, generated by the plasma current, is typically about ten times smaller than B_ϕ. The resulting magnetic field lines are helical and lie on magnetic flux surfaces which are approximately circular in cross-section. In general, a given magnetic field line will completely cover the surface (after an infinite number of trips around the torus). However, there are some surfaces, known as rational flux surfaces, where a field line will form a closed loop after one or more trips around the torus.

The maximum plasma current in a tokamak is limited by stability considerations. In order to avoid the most unstable modes, the "safety factor" $q(a) = aB_\phi/(R_oB_\theta(a))$ must be greater than about 2.5. Consequently, there is a limit to the maximum ohmic heating power one can apply to the plasma. The problem is compounded by the fact that the plasma resistivity drops as the temperature rises. Additional heating is needed to obtain the temperatures required for controlled fusion. The main additional heating methods involve injection of high energy neutral atoms and/or radio frequency heating.

It is not possible, in high temperature laboratory devices, to locate wave launching antennas inside the plasma. Consequently, a basic objective of all wave heating schemes is to heat the plasma from the inside out. If plasmas behaved like metallic conductors, wave energy would be absorbed within a few skin depths in the surface. However, magnetised plasmas do not behave like metallic conductors, and will support a variety of wave types which can propagate, without significant attenuation, from the plasma edge to the centre. These waves will, however, deposit energy in the plasma if there is a resonant surface located somewhere between the centre and the edge. All wave heating schemes rely on the absorption of wave energy at such a surface. The resonance may be due to cyclotron motion of the particles, or to collective particle motions (eg. at the lower or upper hybrid resonance frequencies or at an ion–ion hybrid resonance frequency) or to mode conversion of one wave type to another. In all cases, resonance conditions are encountered as a result of variations in the steady magnetic field or the particle density over the plasma cross–section.

The *Alfven wave heating scheme* involves low frequency (~2MHz) generation of a compressional Alfven wave in an inhomogeneous plasma, using an antenna located at the plasma edge. At frequencies below the ion cyclotron frequency, the compressional wave will mode convert to a torsional Alfven wave at a particular magnetic flux surface in the plasma, known as the Alfven resonance surface, where both wave types locally have the same wavelength in a direction parallel to the steady magnetic field. Since the torsional wave is magnetically guided along steady field lines, wave energy accumulates at this surface. Electrons in and near the surface are locally heated by Landau damping of the torsional wave since the phase velocity of the torsional wave is comparable with the electron thermal speed for typical laboratory plasmas. By appropriate choice of wave frequency and mode numbers, the Alfven resonance surface can be located at any desired position between the plasma centre and the plasma edge (see Fig. 1.2).

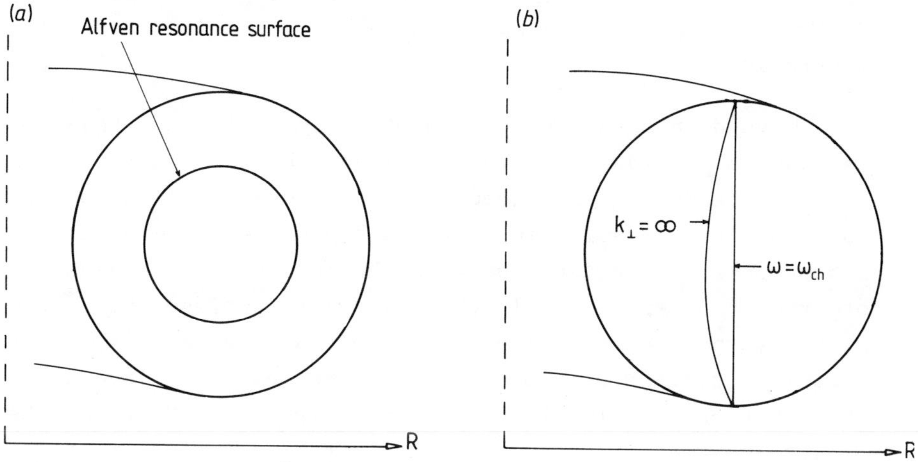

Fig. 1.2 Resonance surfaces in a toroidal plasma. R is the major radius coordinate. The steady (toroidal) magnetic field is proportional to 1/R and the density profile is assumed to be parabolic. (a) Alfven resonance surface in a single ion species plasma. (b) the hydrogen cyclotron layer ($\omega = \omega_{ch}$) and the ion–ion hybrid resonance surface ($k_\perp = \infty$) in a plasma containing ~5% hydrogen and ~95% deuterium.

The ICRF heating schemes operate at higher wave frequencies, typically 20 – 200 MHz, near the ion cyclotron frequency or its harmonics. There are four distinctly different ICRF schemes, all of which have achieved very significant heating in large tokamak devices. In three of these schemes, the compressional Alfven wave is used to carry wave energy from an external antenna to the plasma interior, but it has the wrong (right-hand) polarisation, in a single ion species plasma, to heat ions at the fundamental ion cyclotron frequency. Heating is possible at the *second harmonic*, provided there is a gradient in the wave electric field, so that the acceleration of an ion during one half of its orbit is greater than the deceleration in the other half. This effect is easily understood, at least qualitatively, by considering the motion of single particles. It is not necessary for the wave amplitude to vary with position, since a field gradient, in a direction perpendicular to **B**, will exist provided the wave has a finite perpendicular wavelength.

Wave damping at the fundamental ion cyclotron frequency is possible, however, in a plasma containing two ion species, eg hydrogen and deuterium. As described in Chapter 13, the two species interact to produce an ion-ion hybrid resonance at a frequency between the cyclotron frequencies of the two species. In the *minority heating scheme*, the minority species concentration is only a few percent. The minority cyclotron and ion-ion hybrid resonance layers are then physically very close, as shown in Fig. 1.2, with the result that there is a strong left-hand component of the wave field near the cyclotron layer. The minority species is directly heated at its fundamental cyclotron frequency, and subsequently transfers energy to the majority species and to electrons by Coulomb collisions. The minority species can be accelerated to very high energies in this scheme. In this respect, the scheme has features in common with neutral beam injection heating methods.

In the *wave conversion scheme*, the minority concentration is typically about 20% and the ion - ion hybrid layer is well separated from the two cyclotron layers. In this scheme, the compressional wave mode converts to an ion Bernstein wave at the hybrid resonance layer. Wave heating is due to Landau and cyclotron damping of the Bernstein wave which propagates, both along and across field lines, towards the cyclotron layer of one ion species. The fourth ICRF heating scheme involves *direct excitation of the Bernstein wave* by means of an antenna carrying RF current parallel to the steady field. In each of the other ICRF schemes, the external antennas carry RF current perpendicular to the steady field in order to couple efficiently to the compressional wave. The latter orientation of the antennas is also favoured in the Alfven wave heating scheme.

1.3 Alfven waves in space plasmas

In a short book such as this, it is impossible to do justice to the large amount of work concerned with Alfven waves in space and solid state plasmas. The emphasis in this book is on laboratory plasmas, particularly tokamak plasmas. For the benefit of readers who wish to pursue space physics applications, a short list of references is included in Section M at the end of the book. An excellent starting point is the book by PARKER (M13). The reader will quickly discover that Alfven waves permeate the universe. They have been observed not only in the sun but in the ionosphere and the magnetosphere and even at the earth's surface. Correlation between satellite and ground observations provide direct evidence that the kinetic Alfven wave is guided through the magnetosphere by the earth's magnetic field (M16). There is of course no plasma at the earth's surface, but Alfven wave fields can still be detected in the insulating boundary adjacent to the ionosphere. Alfven wave fields can also be measured in the air space surrounding laboratory plasmas for the same reason. Even if the plasma is surrounded by a metal vessel, the wave fields at the plasma edge can easily penetrate glass covered port openings.

One of the greatest difficulties one encounters in trying to relate Alfven waves in space to laboratory observations is the shocking difference in time and length scales. It is hard to believe that the physics of wave propagation could be the same, but it is. For example, the cyclotron radius of a 20 keV proton at $B \sim 200$nT, is about 200km. Kinetic effects are therefore important in describing an Alfven wave in the magnetosphere if the perpendicular wavelength is smaller than about 1000km. In laboratory plasmas, the corresponding figure is a few cm. The ion cyclotron frequency of a proton, at B ~ 200nT, is 3 Hz. Low frequency Alfven waves in the magnetosphere, or in the sun, are waves with a period of a few minutes. In the laboratory, a low frequency Alfven wave has a period of a few microseconds. Interstellar space is even more remote from laboratory conditions. A low frequency wave in a slightly ionised region where $n_e \sim 1000\ m^{-3}$ may have a period of 10 years, comparable with the particle collision time. The author has been told several times by eminent space physicists that Alfven waves cannot be studied in the laboratory since the wavelengths are much too long! Their mistake was not realising that the wavelength can be as short as one likes, provided the wave frequency and plasma density are sufficiently high. Chosing a heavy gas such as argon also helps (Fig. 6.1) but is not essential. In fact,

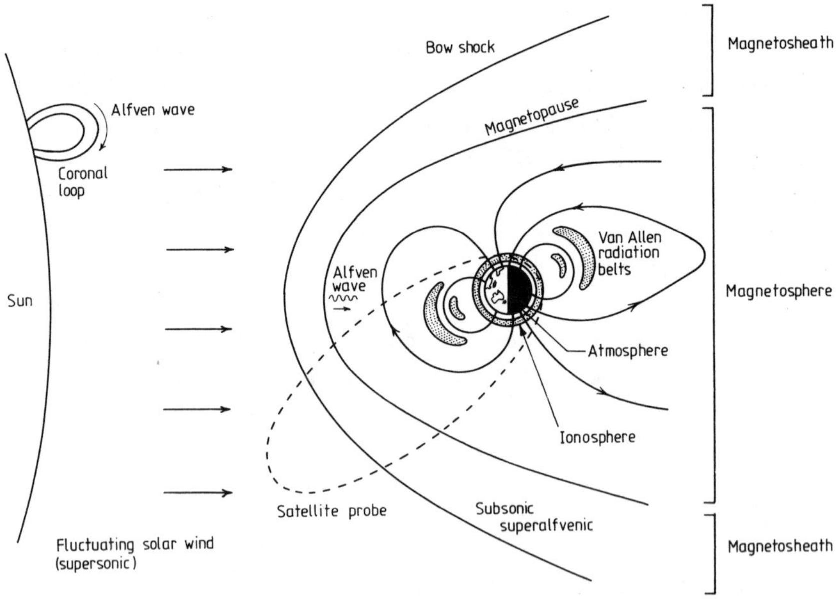

Fig. 1.3 The space physicist's laboratory. Small fluctuations ("micropulsations") in the earth's magnetic field, with period T ~ 1–600 seconds, are due to Alfven waves generated in the magnetosphere (see Section 14.5).

most laboratory studies of Alfven waves are conducted in hydrogen or deuterium. The solar wind and the magnetosphere, probed with spacecraft (Fig. 1.3), also makes a very good laboratory for the the study of Alfven waves.

CESARSKY (M2) and WENTZEL (M20) have reviewed the effects of Alfven wave generation by cosmic rays. The observed isotropy of cosmic rays arriving at the earth indicates either that there is a very homogeneous distribution of sources within and without the Galaxy, or that a mechanism must exist for scattering cosmic rays before they reach the earth. The mechanism most favoured is scattering by Alfven waves which are generated by cosmic rays themselves. Particles moving through a plasma can generate Alfven waves when the particle velocity exceeds the Alfven velocity. This process is essentially the inverse of Landau damping, which is the process whereby waves give up energy to particles moving at about the same speed as the wave. Cosmic rays travelling along a magnetic field line are subsequently scattered by the wave, resulting in an isotropic distribution at the earth. This effect may also be important in laboratory plasmas since alpha particles created in fusion reactions may be scattered by Alfven waves generated by the alpha particles themselves.

In recent years, it has been recognised that Alfven waves may be responsible for heating the solar corona. The surface temperature of the sun is about 6000K. Above the surface, the temperature rises slowly and then jumps through a narrow region to about 1 or 2 $\times 10^6$ K in the solar corona. The fact that the corona is much hotter than the sun's surface has been a long - standing mystery. It is still not well understood, especially since there is good experimental data to maintain checks on the theoretical models.

Most coronal heating models involve transport of wave energy from below the sun's surface and into the corona. Measurements of the flux of acoustic wave energy made in the period 1976–1980 (M10) have shown that it is much too small to heat the corona. Consequently, a strong effort is being made to identify a mechanism to damp Alfven waves in the corona. Several possibilities have been identified, including Alfven resonance heating (M3,M5,M14,M15), phase-mixing of the Alfven wave (G7, M12) and mode conversion to acoustic waves in weak magnetic field regions. Tearing mode turbulence may also be involved (M15). Magnetic flux tubes emerging from the sun pass as gigantic loops through the corona. They are remarkably stable, and may exist for periods of hours or days. A coronal magnetic loop has a structure not unlike a tokamak plasma and a total length of order 100,000 km (M14). Such a loop acts as a resonant cavity for Alfven waves, with a fundamental period ranging from a few seconds to a few minutes, depending on the length of the loop. The theory of Alfven resonance heating, described in Chapter 14, is equally valid for tokamak plasmas and the solar corona. The stability of coronal loops should also be of interest to the fusion community, although the interest tends to be reversed (M14) since there is now much more detailed experimental data available from tokamak devices.

1.4 Theoretical approximations

In most laboratory plasmas, the Alfven wavelengths are comparable with the dimensions of the plasma. Since the plasma parameters vary significantly over a wavelength, most theoretical treatments of wave propagation remain approximate. In this book the simplest possible approximations are made, assuming at first that the plasma is homogeneous and infinite in extent. This approximation may seem to be totally unrealistic but it is a useful first approximation, especially for the torsional Alfven wave. This wave is strongly guided by the steady magnetic field and does not, therefore, reflect off the containing vessel walls. In this respect, the plasma boundary is effectively at infinity.

The basic equations are simplifed throughout most of the text by ignoring plasma kinetic pressure. The kinetic pressure can usually be neglected when the acoustic speed, a, is much less than the Alfven speed, v_A, where

$$a = \left[\frac{2\gamma kT}{m_i} \right]^{1/2} \qquad \text{and} \qquad v_A = \frac{B}{(\mu_0 n_i m_i)^{1/2}}$$

and where T is the plasma temperature, m_i the ion mass, B the steady magnetic field, and n_i is the ion number density. For a hydrogen plasma (m_i = 1.67 x 10^{-27} kg) with T = 10^6 K, n_i = 10^{19} m^{-3}, B = 1T and $\gamma = 5/3$, a = 1.66 x 10^5 m/s and v_A = 6.90 x 10^6 m/s.

These conditions are typical of small tokamak devices. In large tokamaks, and in most other magnetically confined laboratory plasmas, the condition $a \ll v_A$ is usually well satisfied. Nevertheless, kinetic effects are important, even when $a \ll v_A$, if the ion gyroradius is a significant fraction of the perpendicular wavelength. There are circumstances in cold plasma theory where the perpendicular wavelength may be zero (eg at an Alfven resonance layer or at an ion–ion hybrid resonance). Kinetic theory shows that in these cases, the perpendicular wavelength is not zero but it is comparable with the ion gyroradius. These effects are considered in Section 12.7. Effects due to finite plasma resistivity are examined in Chapter 15. Elsewhere the plasma will be regarded as a cold fluid of zero resistivity containing either one or two ion species. The term "cold plasma" means that the temperature is sufficiently low that one can ignore kinetic effects. Even though the resistivity increases when the temperature decreases, the resistivity is often low enough in laboratory plasmas to be ignored. The attenuation length of an Alfven wave in a laboratory plasma at T > 10^6 K, due to the resistivity, is typically greater than 1 km. We will consider wave frequencies in the range $0 < \omega < 10\,\omega_{ci}$ where $\omega_{ci} = Be/m_i$ is the ion cyclotron frequency. For a hydrogen plasma and B = 1 Tesla, $f_{ci} = \omega_{ci}/2\pi$ = 15.25 MHz.

SPACE FILLER

There are a few spaces in this book which could have been left blank but are instead replaced by some food for thought. Here is the first item. The torsional Alfven wave was "discovered" by Hannes Alfven. Who discovered the compressional Alfven wave?

2. THE PROPAGATION VECTOR

In wave theory, the propagation vector, **k**, plays a dominant role. As shown in Fig. 2.1, **k** gives the direction of propagation of plane wavefronts (surfaces of constant phase) and has magnitude $2\pi/\lambda$ where λ is the distance between wavefronts of constant phase.

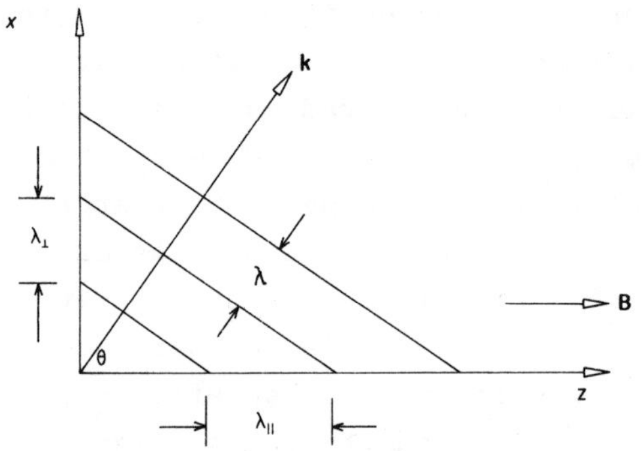

Fig. 2.1 Plane wave propagation vector **k**.

If the magnetic field, **B**, is in the z direction and the propagation vector, **k**, is the the x–z plane at an angle θ to **B**, then **k** has components

$$k_{\parallel} = k\cos\theta = \frac{2\pi}{\lambda_{\parallel}} \qquad k_{\perp} = k\sin\theta = \frac{2\pi}{\lambda_{\perp}} \tag{2.1}$$

The magnitude of **k** is given by $k = (k_{\perp}^2 + k_{\parallel}^2)^{\frac{1}{2}}$. If the phase velocity of the wave in a direction parallel to **k** is $v = f\lambda = \omega/k$, then the phase velocities parallel and perpendicular to **B** are respectively,

$$\left.\begin{aligned} v_{\parallel} &= f\lambda_{\parallel} = \omega/k_{\parallel} = v/\cos\theta \\ v_{\perp} &= f\lambda_{\perp} = \omega/k_{\perp} = v/\sin\theta \end{aligned}\right\} \tag{2.2}$$

The phase velocity, v, is not a vector since the "components" $v_{\parallel}$ and $v_{\perp}$ do not add vectorially. The direction of propagation is best described in terms of the vector **k**, sometimes called simply the wave vector. "High $k_{\perp}$" waves are plane waves propagating at almost 90° to **B** and with small $\lambda_\perp$. The propagation vector does NOT normally give the direction of propagation of wave energy. It gives the direction of energy flow only if the wave is isotropic, i.e., if the phase velocity is independent of θ. The direction of energy flow for the Alfven wave types is discussed in Chapter 4.

Under some conditions, one or more of the components of the propagation vector may have an imaginary part or may be entirely imaginary. If for example **k** has components (k_x, k_y, k_z) and if $k_z = k_R + ik_I$ then the wave fields can be expressed in the form

$$e^{-k_I z}\, e^{i(k_x x + k_y y + k_R z - \omega t)} \tag{2.3}$$

indicating that the wave attenuates exponentially in the z direction. The attenuation length, L, is given by $L = 1/k_I$. In a distance $z = L$, the wave amplitude decreases by a factor 1/e. An imaginary part of k_z arises if a dissipation mechanism, such as resistivity, is included in the plasma equations.

If one of the components of **k** is entirely imaginary, say $k_x = ik_I$ then the wave is evanescent in the x direction, but it may still propagate in the y and/or z directions. The Poynting vector is then perpendicular to the x direction. This situation arises with surface waves, as described in Chapter 8, in waveguides at frequencies below the waveguide cutoff frequency and in inhomogeneous plasmas under conditions where $k_\perp{}^2$ or $k_\parallel{}^2 \to 0$.

In cylindrical geometry, the wave fields can also be described in terms of a propagation vector **k**, having components $k_\perp$ and $k_\parallel$, where $k_\parallel$ is the component parallel to **B** and where $k^2 = k_\perp{}^2 + k_\parallel{}^2$. The radial variation of the wave fields for a homogeneous, cylindrical plasma is given in terms of Bessel functions of the form $J_m(k_\perp r)$, where m is the number of wavelengths in a circumference. However, $k_\perp$ cannot sensibly be decomposed into r and θ components in cylindrical geometry. The basic problem is that the wavelength in the r direction varies with r and is therefore not a well-defined quantity. The wavelength in the θ direction also varies with r, but it does not vary with θ. One can therefore define a local azimuthal wavenumber $k_\theta = 2\pi/\lambda_\theta = m/r$, but there is no equivalent definition of k_r. The wave fields in cylindrical geometry are normally expressed in the form

$$f(r)e^{i(k_z z + k_\theta s - \omega t)} = f(r)e^{i(k_z z + m\theta - \omega t)}$$

where $s = r\theta$ is an arc length and where $k_\theta s = m\theta$ is independent of r.

SPACE FILLER

The diagram below shows the resultant magnetic field lines surrounding a wire carrying a current in a direction perpendicular to a uniform magnetic field. The force, **F**, acting on the wire can be pictured as being due to tension in the field lines. This picture is very satisfying and almost self evident, but is it correct ? Outside the wire there is no $\mathbf{j} \times \mathbf{B}$ force and there is nothing for the force to act on. Note also that the N and S poles experience a force of attraction. Can this force be attributed to tension in the field lines? Is it correct to speak of magnetic field lines as "lines of force"?

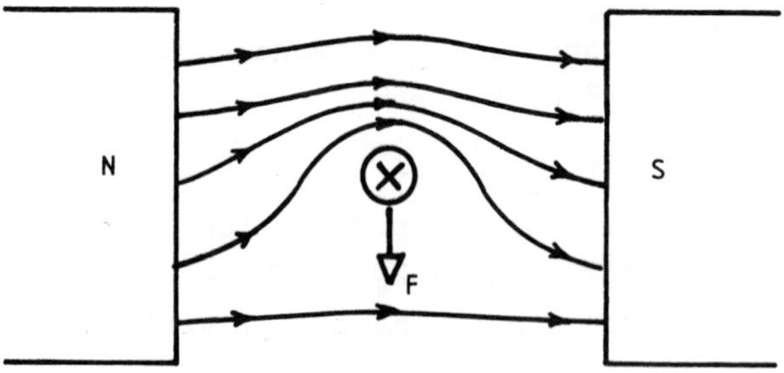

3. LOW FREQUENCY ALFVEN WAVES

Alfven waves can be described at the most basic level in terms of the single fluid, ideal MHD equations and Maxwell's curl equations

$$\rho \frac{\partial \mathbf{v}}{\partial t} = \mathbf{j} \times \mathbf{B} \qquad \mathbf{E} + \mathbf{v} \times \mathbf{B} = 0 \tag{3.1}$$

$$\nabla \times \mathbf{b} = \mu_0 \mathbf{j} \qquad \nabla \times \mathbf{E} = -\frac{\partial \mathbf{b}}{\partial t} \tag{3.2}$$

where ρ is the plasma density, $\mathbf{B}$ is the steady magnetic field, and the quantities $\mathbf{v}$, $\mathbf{j}$, $\mathbf{E}$ and $\mathbf{b}$ are time varying quantities involved in wave propagation. We assume throughout this book that the amplitude of the wave magnetic field, $\mathbf{b}$, is much smaller than the steady magnetic field. Both ρ and $\mathbf{B}$ may vary with position but we assume in this Chapter that they don't.

The relation $\mathbf{E} + \mathbf{v} \times \mathbf{B} = 0$ is a simplified version of Ohm's law for a plasma of zero resistivity, valid at frequencies well below the ion cyclotron frequency. Equations (3.1) are "cold" plasma approximations in which pressure gradients are ignored. Displacement current is ignored in Maxwell's $\nabla \times \mathbf{b}$ equation. This is valid at low wave frequencies except at very low plasma densities where v_A approaches or exceeds the speed of light.

The above set of equations is sufficient to deduce the main properties of Alfven waves at low wave frequencies. Changes in the plasma density due to wave motion can be calculated from the equation of continuity

$$\frac{\partial \rho}{\partial t} = -\nabla\cdot(\rho\mathbf{v}) = -\rho\nabla\cdot\mathbf{v} - \mathbf{v}\cdot\nabla\rho$$

expressing conservation of mass. For small amplitude perturbations, $\mathbf{v}$ and $\nabla\rho$ are both small and their product can be neglected.

Equations (3.1) and (3.2) can be solved either in Cartesian (x,y,z) or cylindrical (r,θ,z) coordinates. Solutions for a cylindrical plasma will be examined in Chapter 9. In this Chapter we examine plane wave solutions of the form

$$e^{i(k_x x + k_y y + k_z z - \omega t)} \tag{3.3}$$

so that $\partial/\partial x = ik_x$, $\partial/\partial y = ik_y$, $\partial/\partial z = ik_z$ and $\partial/\partial t = -i\omega$. It is even simpler to rotate the coordinate system so that **k** lies in the x–z plane ($k_y = 0$) but we consider the case $k_y \neq 0$ for generality. Assuming that **B** is in the z direction, the components of equation (3.1) are

$$\left.\begin{aligned} -i\omega\rho v_x &= j_y B & E_x &= -v_y B \\ i\omega\rho v_y &= j_x B & E_y &= v_x B \\ v_z &= 0 & E_z &= 0 \end{aligned}\right\} \tag{3.4}$$

$$\therefore \quad i\omega\mu_o j_x = \omega^2 E_x/v_A{}^2 \qquad i\omega\mu_o j_y = \omega^2 E_y/v_A{}^2 \tag{3.5}$$

where $v_A{}^2 = B^2/(\mu_o\rho)$. Since $E_z = 0$, the components of (3.2) are

$$\left.\begin{aligned} \mu_o j_x &= i(k_y b_z - k_z b_y) & \omega b_x &= -k_z E_y \\ \mu_o j_y &= i(k_z b_x - k_x b_z) & \omega b_y &= k_z E_x \\ \mu_o j_z &= i(k_x b_y - k_y b_x) & \omega b_z &= k_x E_y - k_y E_x \end{aligned}\right\} \tag{3.6}$$

From equations (3.5) and (3.6),

$$\left.\begin{aligned} (k_y{}^2 + k_z{}^2 - \omega^2/v_A{}^2)E_x &= k_x k_y E_y \\ (k_x{}^2 + k_z{}^2 - \omega^2/v_A{}^2)E_y &= k_x k_y E_x \end{aligned}\right\} \tag{3.7}$$

$$\therefore \quad (k_x{}^2 + k_y{}^2 - \omega^2/v_A{}^2)(k_y{}^2 + k_z{}^2 - \omega^2/v_A{}^2) = k_x{}^2 k_y{}^2$$

or

$$(\omega^2/v_A{}^2 - k_\parallel{}^2)(\omega^2/v_A{}^2 - k^2) = 0 \tag{3.8}$$

where $k_\parallel = k_z$, $k_\perp{}^2 = k_x{}^2 + k_y{}^2$ and $k^2 = k_\perp{}^2 + k_\parallel{}^2$.

The two solutions of Eq.(3.8) are

$$\boxed{\omega/k_\parallel = v_A} \qquad \text{TORSIONAL WAVE} \tag{3.9}$$

and $\boxed{\omega/k = v_A}$ COMPRESSIONAL WAVE (3.10)

Using these solutions, we can find the electric field components from Eq.(3.7), the magnetic field components from Eq.(3.6), and the **v** components from Eq.(3.4), as follows:

TORSIONAL	COMPRESSIONAL	
$E_y/E_x = k_y/k_x$	$E_y/E_x = -k_x/k_y$	
$E_z = 0$	$E_z = 0$	
$b_y/b_x = -k_x/k_y$	$b_y/b_x = k_y/k_x$	
$b_z = 0$	$b_z/b_y = -k_\perp{}^2/(k_y k_\parallel)$	(3.11)
$\nabla\cdot\mathbf{E} = i\mathbf{k}\cdot\mathbf{E} \neq 0$	$\nabla\cdot\mathbf{E} = 0$	
$\nabla\cdot\mathbf{v} = i\mathbf{k}\cdot\mathbf{v} = 0$	$\nabla\cdot\mathbf{v} = i\omega b_z/B$	
$\nabla\cdot\mathbf{b} = i\mathbf{k}\cdot\mathbf{b} = 0$	$\nabla\cdot\mathbf{b} = i\mathbf{k}\cdot\mathbf{b} = 0$	

The phase velocity of the torsional wave in a direction parallel to **B** is equal to v_A, regardless of the direction of **k**. The phase velocity of the compressional wave is equal to v_A, regardless of the direction of **k**. If θ is the angle between **k** and **B**, then $\omega/k = v_A \cos\theta$ for the torsional wave and $\omega/k = v_A$ for the compressional wave.

The torsional wave is partly electrostatic, resulting in charge separation. Purely electrostatic waves have **E** parallel to **k**, so $\nabla \times \mathbf{E} = i\mathbf{k} \times \mathbf{E} = i\omega\mathbf{b} = 0$. The torsional wave does not involve any compression of the plasma since $\nabla\cdot\mathbf{v} = 0$. This makes it hard to observe the torsional wave by laser scattering off density fluctuations. However, small density fluctuations can be observed, due to finite frequency or kinetic effects. The compressional wave, as its name implies, acts to compress the plasma but there is no charge separation. The torsional wave is transverse magnetic ($b_z = 0$) at low wave frequencies although it does develop a significant b_z component when $\omega \gtrsim 0.3\ \omega_{ci}$. The compressional wave has a strong b_z component, except for the special case $k_\perp = 0$ where the propagation vector is parallel to **B** and then $b_z = 0$. If **k** is perpendicular to **B** ($k_\parallel = 0$) then b_z is the only magnetic field component for the compressional wave.

The relative directions of the wave fields can be found from (3.11), with $\mathbf{k}\cdot\mathbf{E} = k_xE_x + k_yE_y + k_zE_z$ etc. as follows:

$$\left.\begin{array}{lll} \underline{\text{TORSIONAL}} & & \underline{\text{COMPRESSIONAL}} \\ \mathbf{k}\cdot\mathbf{E} \neq 0 & & \mathbf{k}\cdot\mathbf{E} = 0 \\ \mathbf{k}\cdot\mathbf{b} = 0 & & \mathbf{k}\cdot\mathbf{b} = 0 \\ \mathbf{k}\cdot\mathbf{v} = 0 & & \mathbf{k}\cdot\mathbf{v} \neq 0 \\ \mathbf{k}\cdot\mathbf{j} = 0 & & \mathbf{k}\cdot\mathbf{j} = 0 \\ \mathbf{E}\cdot\mathbf{b} = 0 & & \mathbf{E}\cdot\mathbf{b} = 0 \end{array}\right\} \quad (3.12)$$

These relations can be visualised more clearly if we rotate the coordinate system so that $k_y = 0$. Then $\mathbf{k}$ lies in the x–z plane as shown in Fig. 3.1 and the wave field components are ;

$$\left.\begin{array}{lll} \underline{\text{TORSIONAL}} & & \underline{\text{COMPRESSIONAL}} \\ E_y = E_z = 0 & & E_x = E_z = 0 \\ b_x = b_z = 0 & & b_y = 0 \\ j_y = 0 & & j_x = j_z = 0 \\ v_x = v_z = 0 & & v_y = v_z = 0 \end{array}\right\} \quad (3.13)$$

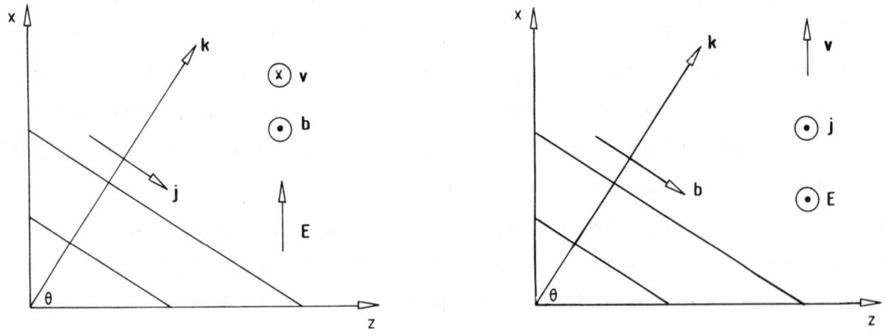

Fig. 3.1 Relative directions of wave fields when $k_y = 0$ and $\mathbf{B}$ is in the z direction.

4. POYNTING AND GROUP VELOCITY VECTORS

In a coordinate system where $k_y = 0$ the wave field components and energy densities of the two Alfven waves are:

TORSIONAL	COMPRESSIONAL	
$E_x \neq 0$	$E_y \neq 0$	
$b_y = E_x/v_A$	$b_x = -k_z E_y/\omega$	
$b_z = 0$	$b_z = k_x E_y/\omega$	
$v_y = -E_x/B$	$v_x = E_y/B$	(4.1)
$W_E = \epsilon_o E_x{}^2/2$	$W_E = \epsilon_o E_y{}^2/2$	
$W_B = b_y{}^2/2\mu_o = c^2 W_E/v_A{}^2$	$W_B = c^2 W_E/v_A{}^2$	
$W_K = \rho v_y{}^2/2 = W_B$	$W_K = W_B$	

where c is the speed of light, W_E and W_B are the energy densities in the electric and magnetic fields respectively and W_K is the kinetic energy density of the plasma. Under the usual laboratory conditions where $v_A \ll c$, the total wave energy is distributed equally between the magnetic and kinetic energies. These components are in phase. Wave energy does not alternate between the W_B and W_K contributions. The total wave energy density is therefore $W_T = b^2/\mu_o$. In the more general case where $k_y \neq 0$ we obtain the same result, and $b^2 = b_x{}^2 + b_y{}^2 + b_z{}^2$.

The direction and magnitude of the instantaneous flux of electromagnetic energy through a surface, in watts/m², is given by the Poynting vector

$$\mathbf{S} = \mathbf{E} \times \mathbf{b}/\mu_o \qquad (4.2)$$

When $E_z = 0$, $\quad \mu_o \mathbf{S} = \hat{\mathbf{x}}(E_y b_z) - \hat{\mathbf{y}}(E_x b_z) + \hat{\mathbf{z}}(E_x b_y - E_y b_x)$

For the torsional wave, we find from equation (3.6) that

$$\mathbf{S}_T = (b^2 v_A/\mu_o)\hat{\mathbf{z}}$$

The Poynting vector for the torsional wave is therefore parallel to **B**, regardless of the direction of **k**. For the compressional wave,

$$\mathbf{S}_C = (\omega b_z{}^2/\mu_0 k_\perp{}^2)\ (\hat{\mathbf{x}}\, k_x + \hat{\mathbf{y}}\, k_y + \hat{\mathbf{z}}\, k_z)$$

The Poynting vector for the compressional wave is therefore parallel to **k**. The magnitude of $\mathbf{S}_C$ is $b^2 v_A/\mu_0$. In both cases we see that S = energy density × v_A. An alternative quantity which can be used to describe the speed and direction of energy flow is the group velocity vector defined by

$$\boxed{\mathbf{v}_g = \frac{\partial\omega}{\partial\mathbf{k}} = \hat{\mathbf{x}}\frac{\partial\omega}{\partial k_x} + \hat{\mathbf{y}}\frac{\partial\omega}{\partial k_y} + \hat{\mathbf{z}}\frac{\partial\omega}{\partial k_z}} \tag{4.3}$$

For the torsional wave, $\omega = k_z v_A$ so $\mathbf{v}_g = \hat{\mathbf{z}}\, v_A$ as expected from the Poynting vector result. For the compressional wave

$$\omega = kv_A = [k_x{}^2 + k_y{}^2 + k_z{}^2]^{1/2}\, v_A$$

$$\therefore \qquad \mathbf{v}_g = (v_A{}^2/\omega)(\hat{\mathbf{x}}\, k_x + \hat{\mathbf{y}}\, k_y + \hat{\mathbf{z}}\, k_z)$$

again illustrating that wave energy in this case propagates in a direction parallel to **k**. In this and other respects, the compressional wave behaves like an electromagnetic wave in free space, the main difference being that the phase velocity is v_A rather than c.

These results are quite straightforward. For some wave types **S** and $\mathbf{v}_g$ are not parallel. For example, a sound wave in air has zero **S** and non-zero $\mathbf{v}_g$. The Poynting vector for the slow MHD (ion-acoustic) wave is almost perpendicular to **B**, but the group velocity vector is essentially parallel to **B**. The Poynting vector does not include thermal energy, but $\mathbf{v}_g$ includes all forms of wave energy and therefore provides a more reliable estimate of the speed and direction of energy flow in general. It should be noted that the group velocity is NOT given by $\partial\omega/\partial k$ in general. For example, $\omega = kv_A\cos\theta$ for the torsional wave, so $\partial\omega/\partial k = v_A\cos\theta$, whereas the magnitude of $\mathbf{v}_g$ is v_A. The quantity $\partial\omega/\partial k$ is the component of $\mathbf{v}_g$ in the direction of **k** (WALKER G11), so $v_g = \partial\omega/\partial k$ only if $\mathbf{v}_g$ is parallel to **k**.

5. PHASE AND GROUP VELOCITY SURFACES

A phase velocity surface (also known as a wave normal surface) represents a polar plot of the phase velocity vs the angle θ between **k** and **B**. For the torsional wave, $\omega/k = v_A \cos\theta$ and for the compressional wave, $\omega/k = v_A$.

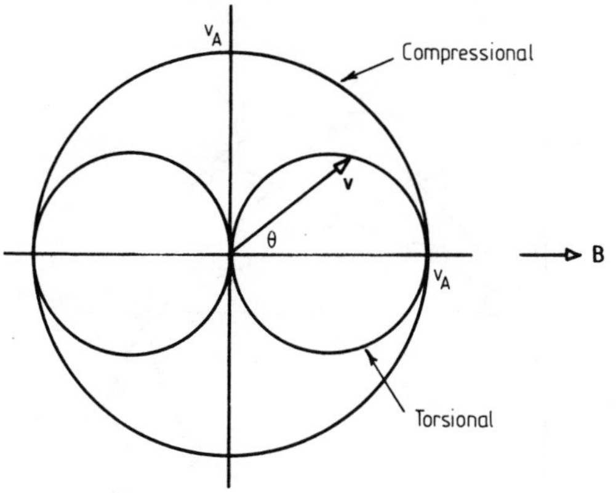

Fig. 5.1 Phase velocity surfaces for Alfven waves in the ideal MHD limit $\omega \ll \omega_{ci}$.

The surface is constructed so that the vector **v** is parallel to **k** and its length is proportional to the phase velocity. The phase velocity itself is not a vector quantity since the velocity "components" $v_{\parallel} = \omega/k_{\parallel}$ and $v_{\perp} = \omega/k_{\perp}$ do not add vectorially. Phase and group velocity surfaces are three dimensional, but are always drawn in projection in a plane containing **B**.

For the isotropic compressional wave the phase velocity surface represents the shape of a wave front for a point source at the origin. It also represents the group velocity surface. However, the phase velocity surface for the torsional wave does <u>not</u> represent the shape of the wave front. This effect can be seen geometrically by constructing plane wave fronts originating at t = 0 from the origin. Three such wave fronts are shown in Fig. 5.2 at t = 1 sec where the fronts have advanced a distance equal to the phase velocity $v_p = v_A \cos\theta$. It will be noted that all wave fronts interfere constructively at the single point P on the steady magnetic field line passing through the source at the origin. There is a similar point at θ = 180°. Consequently, energy in the torsional wave propagates only along field lines as a ray. This method of geometric wave front construction is described in more detail by STIX (A8) and results for other wave types are given by BOOKER (A1), MUSIELAK (G8) and WALKER (G11).

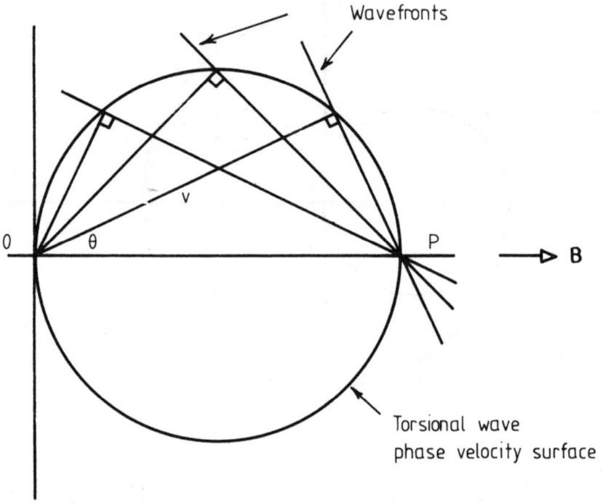

Fig. 5.2 Interference of torsional wave fronts at point P. Wave energy propagates only along magnetic field lines.

A torsional Alfven ray, generated by a point source at the origin, will contain a continuous spectrum of $k_\perp$ components. Each plane wave component propagates with the same phase velocity along **B** so the ray remains coherent. If the source is turned on and off, it will generate a finite length wave packet of zero transverse width. The properties of torsional Alfven rays are quite different from free space rays. A torsional ray contains a continuous spectrum of $k_\perp$ components and a unique $k_\parallel$ component. A narrow beam of light such as a laser beam contains a continuous spectrum of both $k_\perp$ and $k_\parallel$ components since each plane wave component propagates at the same speed (the speed of light) in the direction of **k**. Hence each $k_\perp$ component is associated with a different $k_\parallel$ component. Furthermore, it is not possible to generate a free-space ray whose transverse width is less than one wavelength. If a beam of light is passed through a narrow slit in an attempt to reduce its width, the beam will be diffracted. In optics, the term "ray" denotes a fictitious line of zero width drawn to represent the path of a narrow beam of light. Torsional Alfven rays, on the other hand, can have a physical width equal to zero in the ideal MHD limit.

These properties are derived from the ideal MHD equations. Non-ideal effects such as finite resistivity, finite frequency (compared with ω_{ci}), finite m_e or finite T_e are considered in the following Chapters. It is shown that the phase velocity of the torsional wave does depend slightly on $k_\perp$ for realistic plasma conditions. This gives rise to a finite perpendicular component of the group velocity vector, but it always remains small compared to the parallel component. For example, Fig. 5.3 shows the effect of finite frequency on the phase and group velocity surfaces for the Alfven wave types. The phase velocity surfaces were computed from the dispersion relation Eq.(10.7), expressed in the form

$$\frac{\omega^2}{k^2} = \frac{2\, v_A^{\,2}\,(1 - \Omega_i^{\,2})\cos^2\theta}{1 + \cos^2\theta \pm (\sin^4\theta + 4\Omega_i^{\,2}\cos^2\theta)^{1/2}} \tag{5.1}$$

where the + sign describes the torsional wave and the − sign describes the compressional wave. At any given ω and θ, (5.1) gives the phase velocity ω/k and the components $k_\perp$ and $k_\parallel$. The group velocity surface coordinates $(\partial\omega/\partial k_\parallel, \partial\omega/\partial k_\perp)$ were computed from the formulae in Sect. 10.4. As shown in Fig. 5.3, the group velocity surface for the torsional wave has a peculiar arrowhead shape at finite frequency. The arrowhead shrinks to a point as $\Omega_i \to 0$ and as $\Omega_i \to 1$.

Group velocity surfaces for the compressional wave at high wave frequencies are shown in Fig. 5.4. These surfaces were computed from Eq.(12.11) in order to extend the calculations to frequencies well above the ion cyclotron frequency. As the wave frequency is increased, the compressional wave evolves into a whistler wave and the group velocity surfaces elongate in a direction parallel to the steady B field. These surfaces develop an arrowhead or cusp

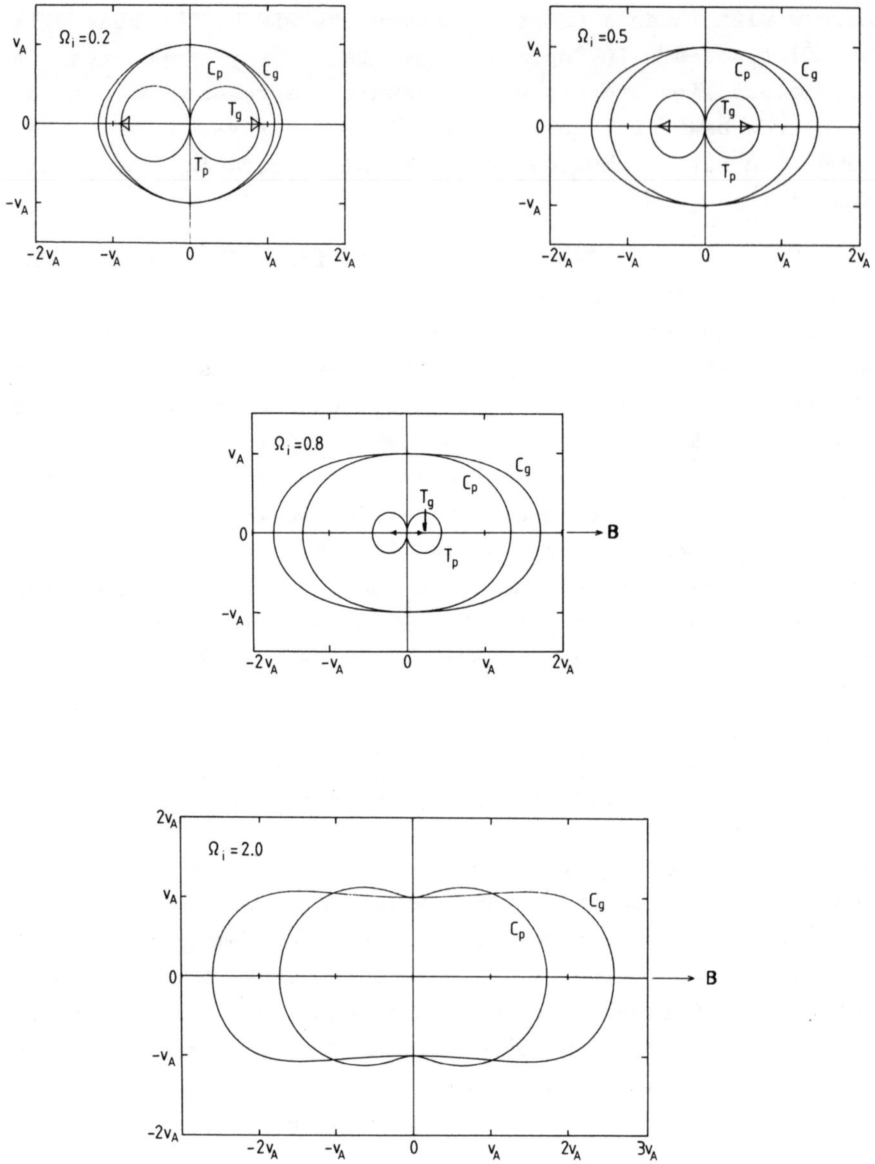

Fig. 5.3 Phase and group velocity surfaces for torsional and compressional waves at different wave frequencies, $\Omega_i = \omega/\omega_{ci}$.

T_p = torsional wave phase velocity surface
T_g = torsional wave group velocity surface
C_p = compressional wave phase velocity surface
C_g = compressional wave group velocity surface

shape when $\Omega_i \gtrsim 40$. The arrowhead subtends an angle of about 19° to the origin. Consequently, wave energy in the whistler wave is confined to propagate in a narrow cone along the steady field lines. The wave is dispersive, giving rise to a characteristic whistling tone when the wave propagates through the ionosphere. By comparison, the torsional wave group velocity surface has a much smaller angular extent, approaching zero as $\omega \to 0$.

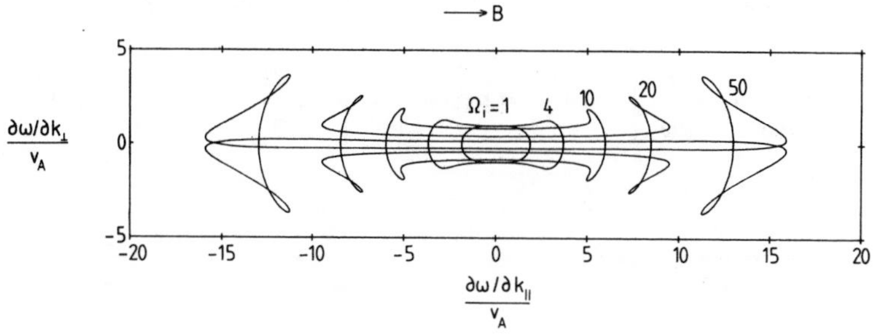

Fig 5.4 Group velocity surfaces for the compressional wave at high wave frequencies, in hydrogen, with $v_A/c = 0.02$.

The equations used to plot the results in Fig. 5.4 are of general interest, since they include all cold plasma wave types and cases where there is more than one ion species. Phase velocity surfaces can be calculated from Eq.(12.10), which is quadratic in n^2, so

$$\frac{v_p{}^2}{c^2} = \frac{1}{n^2} = \frac{B_1 \pm [\, B_1{}^2 - 4A_1C_1]^{\frac{1}{2}}}{2C_1} \tag{5.2}$$

where A_1 and B_1 are given in terms of θ in Chapter 12. The coordinates for the group velocity surface are obtained by differentiating Eq.(12.11) with respect to $k_\parallel$ and $k_\perp$:

$$2k_\parallel(Q + 2Pk_\parallel{}^2) + \frac{\partial\omega}{\partial k_\parallel}\left[\, k_\parallel{}^4\,\frac{\partial P}{\partial\omega} + k_\parallel{}^2\,\frac{\partial Q}{\partial\omega} + \frac{\partial M}{\partial\omega}\right] = 0 \tag{5.3}$$

$$\frac{\partial M}{\partial k_\perp} + k_\parallel{}^2 \frac{\partial Q}{\partial k_\perp} + \frac{\partial \omega}{\partial k_\perp}\left[k_\parallel{}^4 \frac{\partial P}{\partial \omega} + k_\parallel{}^2 \frac{\partial Q}{\partial \omega} + \frac{\partial M}{\partial \omega}\right] = 0 \qquad (5.4)$$

where $\dfrac{\partial M}{\partial k_\perp} = 4Sk_\perp{}^3 - \dfrac{2\omega^2}{c^2}(LR + SP)k_\perp$, $\dfrac{\partial Q}{\partial k_\perp} = 2(P + S)k_\perp$

$$\frac{\partial Q}{\partial \omega} = k_\perp{}^2 \left[\frac{\partial P}{\partial \omega} + \frac{\partial S}{\partial \omega}\right] - \frac{4\omega}{c^2} PS - \frac{2\omega^2}{c^2}\left[P\frac{\partial S}{\partial \omega} + S\frac{\partial P}{\partial \omega}\right]$$

$$\frac{\partial M}{\partial \omega} = k_\perp{}^4 \frac{\partial S}{\partial \omega} - \frac{2\omega k_\perp{}^2}{c^2}(LR + SP) - \frac{\omega^2 k_\perp{}^2}{c^2}\left[L\frac{\partial R}{\partial \omega} + R\frac{\partial L}{\partial \omega} + S\frac{\partial P}{\partial \omega} + P\frac{\partial S}{\partial \omega}\right] + \frac{4\omega^3}{c^4} LRP + \frac{\omega^4}{c^4}\left[LR\frac{\partial P}{\partial \omega} + LP\frac{\partial R}{\partial \omega} + RP\frac{\partial L}{\partial \omega}\right]$$

$$\frac{\partial P}{\partial \omega} = \frac{2}{\omega^3}\Sigma\,\omega_p{}^2 \qquad \frac{\partial S}{\partial \omega} = 2\omega\,\Sigma\frac{\omega_p{}^2}{[\omega^2 - \omega_c{}^2]^2}$$

$$\frac{\partial L}{\partial \omega} = \Sigma\frac{\omega_p{}^2(2\omega - \omega_c)}{[\omega^2 - \omega\omega_c]^2} \qquad \frac{\partial R}{\partial \omega} = \Sigma\frac{\omega_p{}^2(2\omega + \omega_c)}{[\omega^2 + \omega\omega_c]^2}$$

The summation extends over all charged species, including electrons. The sign of ω_c is important, as described in Chapter 12, being negative for electrons, ie $\omega_{ce} = - Be/m_e$. Although these equations look complicated, they are easy to program on a personal computer. For any given value of θ, $k_\perp$ and $k_\parallel$ can be calculated from (5.2). The coordinates $\partial\omega/\partial k_\perp$ and $\partial\omega/\partial k_\parallel$ can then be found from (5.3) and (5.4). These coordinates define the tip of the group velocity vector, both in magnitude and direction. The angle between $\mathbf{v}_g$ and $\mathbf{B}$ so obtained, is not equal to θ in general. Results for a plasma with two ion species are given in Chapter 13.

SPACE FILLER

What is the significance of the Poynting vector in the space around an isolated charge placed in a steady magnetic field? Consider the surface integral of the Poynting flux over a closed surface.

6 EXPERIMENTAL OBSERVATIONS OF ALFVEN WAVES

The first observations of Alfven waves were made over a hundred years ago, but were not recognised as such at the time (M9). They were observed as low frequency fluctuations in the Earth's magnetic field, now known as ULF pulsations or as geomagnetic micropulsations (M6). Modern observations using arrays of ground-based magnetometers and observations made on board satellites in geosynchronous orbit have established the existence of almost monochromatic standing torsional Alfven waves on the Earth's magnetic field lines. Satellites which are in motion at high velocity relative to the background plasma are less suited for this type of measurement. The generation mechanisms are still not fully understood and more extensive observations will be required to identify the excitation sources (M1, M19).

The first observations of Alfven waves in the laboratory were made in mercury (C10) and liquid sodium (C9). Much better observations were subsequently made in the late 1950's (C1,C5,C14,C15), in relatively cold ($T_e \sim$ 1eV), and relatively dense ($n_e \sim 10^{21}$ m^{-3}) hydrogen or argon plasmas. With B ~ 1 Tesla and f ~ 1 MHz in hydrogen, the Alfven wavelength was ~ 1m. The attenuation length, due mainly to resistive damping, was also ~ 1m in these experiments. Early experiments were concerned mainly with measurements of the dispersion relations for torsional and compressional waves. Heating experiments were often disappointing due to the poor energy confinement of the plasma sources. The situation changed dramatically in the 1970's when tokamak devices suddenly appeared on the scene. There are now about 20 tokamak devices in the world which use RF heating at power levels of 1 - 20 MWatt for supplementary heating to conditions where thermonuclear neutrons are generated.

As described in Chapter 17, there are several techniques available to detect Alfven waves in a laboratory plasma. In high temperature devices, these include laser scattering off density fluctuations or, less directly, by measurements of antenna impedance. The simplest technique, and one which

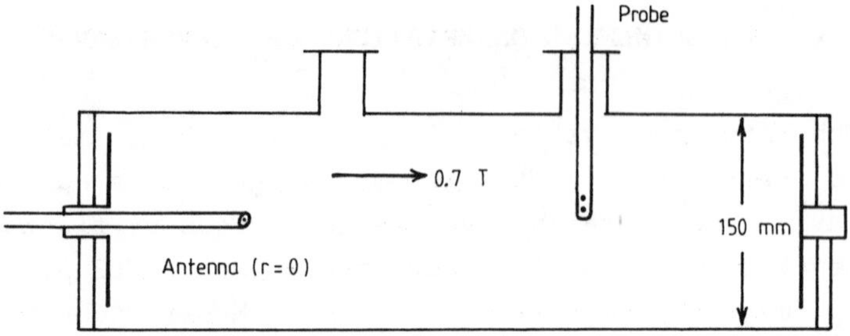

Experimental arrangement

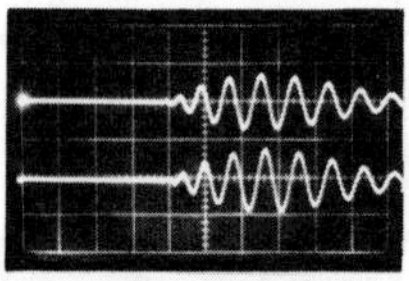

r = 20 mm
r = 5 mm

Compressional Alfven wave in argon ($2\mu s$/div.)

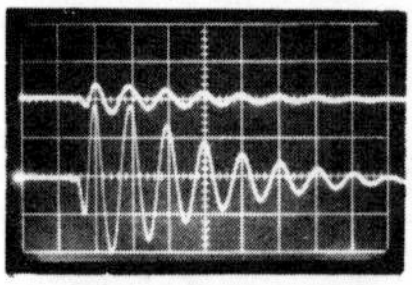

r = 20 mm
r = 5 mm

Torsional Alfven wave in hydrogen ($2\mu s$/div.)

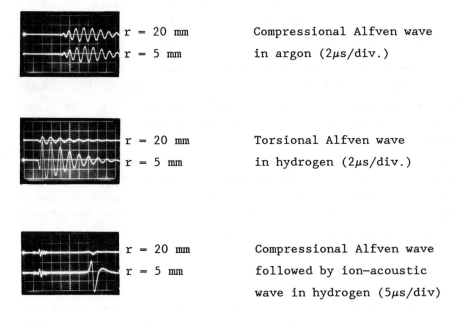

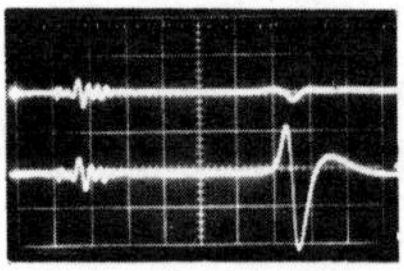

r = 20 mm
r = 5 mm

Compressional Alfven wave followed by ion-acoustic wave in hydrogen ($5\mu s$/div)

Fig. 6.1 Demonstration of wave guidance along magnetic field lines for the torsional and ion-acoustic waves. The compressional wave is essentially isotropic. The time delay from the beginning of the trace yields the Alfven speed.

gives a local measurement of the wave magnetic field, is to use a magnetic probe. This technique is suitable only for measurements in low temperature plasmas or at the edge of a high temperature plasma.

Results showing the essential differences between the Alfven and acoustic wave types are given in Fig. 6.1. In this experiment (G3), a small antenna was inserted in a low temperature (1.5 eV) plasma with an axial magnetic field B = 0.7 T. The antenna was driven by a damped, oscillating current pulse and the wave magnetic fields were measured downstream of the antenna. Due to strong resistive damping of each wave type, there were no reflections from the far end of the device to interfere with the wave fields between the antenna and the probe. The acoustic wave was generated using a small spark gap to create a pressure disturbance. The torsional wave was generated in a hydrogen plasma at a frequency below the cutoff frequency for the compressional wave so that it could be observed without interference from the compressional wave. The compressional wave alone was observed in an argon plasma at a frequency above the ion cyclotron frequency.

As shown in Fig. 6.1, the torsional wave was guided by the steady magnetic field and was observed a short distance away from the antenna only on field lines passing through or near the antenna. Further downstream, the wave was observed to diffuse across field lines due to strong resistive damping of the shorter perpendicular wavelength components. In a tokamak plasma, diffusion is much weaker due to the lower resistivity and the torsional wave is magnetically guided for a much greater distance (see Figs. 6.3 – 6.6).

The compressional wave propagated isotropically away from the antenna and was observed with similar amplitude both across **B** lines and along **B** lines. The acoustic wave was strongly guided by the steady field, as expected, since its group velocity is also directed parallel to **B**. The acoustic and torsional Alfven wave modes are the only plasma waves with a group velocity vector almost exactly parallel to **B**. Even the whistler wave, which is guided through the ionosphere from one hemisphere to the next, has a significant group velocity component perpendicular to **B** (see Fig. 5.4). The whistler wave is guided primarily as a result of density gradients in the ionosphere rather than by the **B** field alone. Wave guidance by a density gradient is due to reflection and/or refraction and is known as ducting.

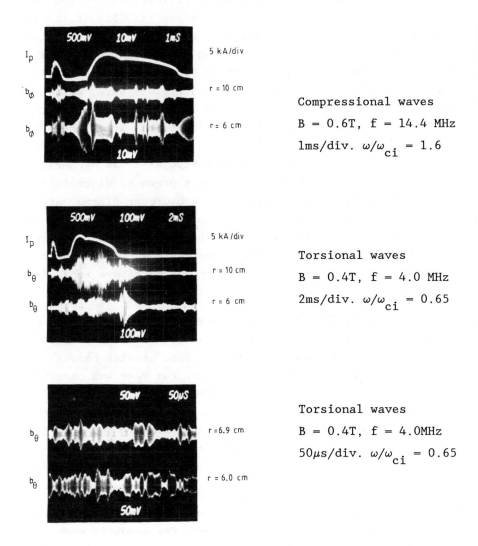

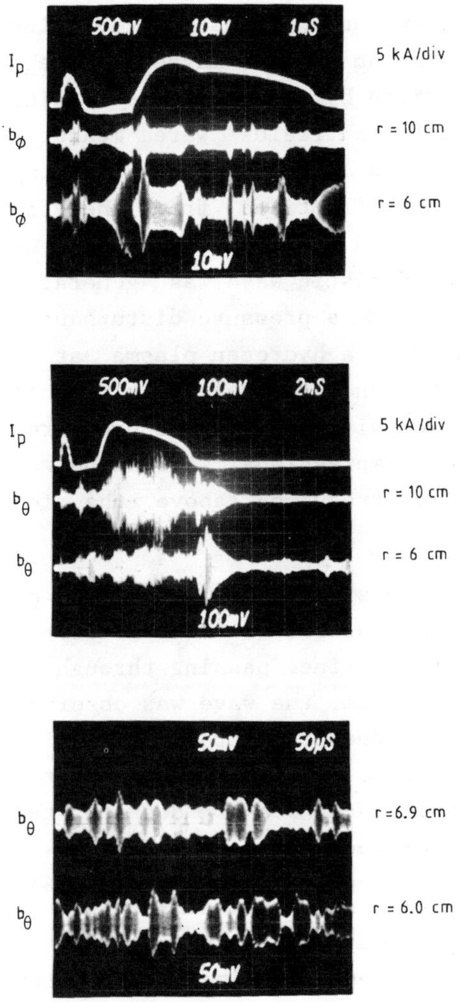

Compressional waves
B = 0.6T, f = 14.4 MHz
1ms/div. $\omega/\omega_{ci} = 1.6$

Torsional waves
B = 0.4T, f = 4.0 MHz
2ms/div. $\omega/\omega_{ci} = 0.65$

Torsional waves
B = 0.4T, f = 4.0MHz
50μs/div. $\omega/\omega_{ci} = 0.65$

Fig. 6.2 Alfven wave magnetic field signals in a tokamak hydrogen plasma, minor radius 11cm, measured simultaneously with two probe coils at radii r.

Differences between torsional and compressional wave propagation in a tokamak plasma are indicated in Fig. 6.2 (from Ref. F5). This figure shows typical wave magnetic field signals measured inside the plasma, excited by a sinusoidal (CW) current through a full-turn loop antenna surrounding the plasma (as in Fig. 17.1). At high wave frequencies, above the ion cyclotron frequency, only the compressional wave can propagate. Sharp peaks in the wave signals are due to cavity modes which pass into and out of resonance due to the variation with time of the plasma density. The resonance peaks are observed simultaneously at all points in the plasma as well as near the plasma edge, indicating that the resonances are true cavity modes.

At low wave frequencies, below the ion cyclotron frequency, the wave magnetic field signals are dramatically different in character. Except for a narrow frequency range associated with discrete Alfven modes (to be described in Sect. 11.3), there are no cavity modes and the magnetic field waveforms measured at any two points in the plasma cross-section, even 9mm apart, are quite different. As shown in Fig. 6.2, the wave amplitude varies rapidly on a time scale 20–50μs, which is the time scale for local variations in the plasma density due to MHD activity. Consequently, the time variation in the Alfven wave fields can be attributed to MHD activity in the plasma, but a precise mechanism has not yet been found to account for the structure of the wave fields.

It is clear from these results that wave energy is propagated along field lines by the torsional wave, and not across field lines. Strong amplitude modulation of the wave fields indicates interference of two or more waves which have arrived at the probe location by different routes or at different speeds. The amplitude modulation seen in Fig. 6.2 could therefore be due either to interference of torsional and compressional waves or to torsional waves travelling along field lines in opposite directions around the tokamak. It is difficult to interpret these results since it is difficult to measure the electron density variation along a given helical field line, or even at two points on the same field line. Not only does the density fluctuate locally with time by up to 50%, but the field lines themselves wander about as a result of the magnetic field perturbations associated with MHD modes. A field line shift of 10mm over a distance of 1 metre can be generated by a transverse magnetic field perturbation of only 1% of the toroidal field.

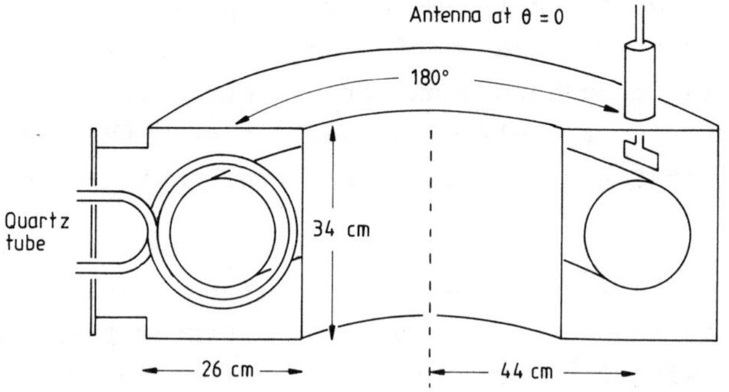

Fig. 6.3 Experimental arrangement used to observe guided ray propagation in a tokamak plasma.

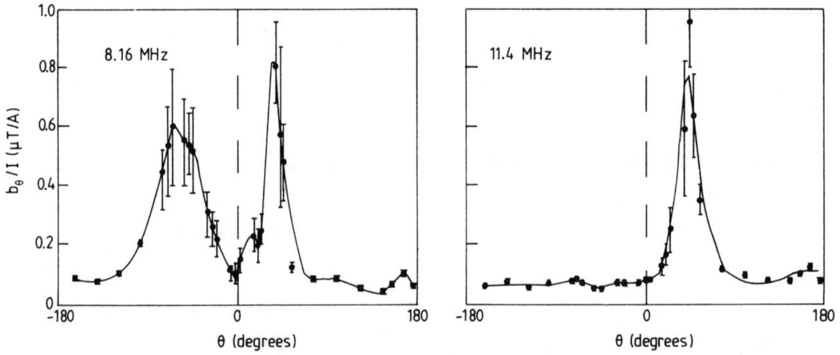

Fig. 6.4 Results demonstrating wave guidance in a tokamak hydrogen plasma with B = 0.71T and f_{ci} = 10.8 MHz at θ = 0. The wave magnetic field, b_θ, is normalised to the antenna current, I. The angle θ is defined in Fig. 6.6.

More direct evidence that the torsional wave is magnetically guided along curved magnetic field lines is given in Figs. 6.3 to 6.5. These results demonstrate that the wave is guided even when the steady field lines are helically twisted and even at frequencies up to the ion cyclotron frequency. Furthermore, the results show that the wave remains guided in the presence of a strong density gradient at the plasma edge, and it is guided at low wave frequencies even though the parallel wavelength is much longer than the plasma circumference. In this experiment (G2), a localised disturbance was generated by a small loop antenna located at the edge of a tokamak plasma. The ray trajectory was determined with an array of magnetic probes at the plasma edge. The probes were housed in a quartz tube, surrounding the plasma poloidally, and oriented to detect the poloidal (b_θ) component of the wave magnetic field. The profile of b_θ vs θ shows two distinct peaks, corresponding to ray paths from the antenna to the high and low field sides of the torus. Further transits of these rays around the torus were blocked by vertical limiters. As shown in Fig. 6.4, the ray on the low field side disappeared when the wave frequency was increased so that it was slightly above the ion cyclotron frequency on the low field side of the torus. The ray on the high field side continued to propagate since the wave frequency remained below ω_{ci} on the high field side.

The same arrangement shown in Fig. 6.3 was also used to obtain the results in Fig. 6.5. These results (G2) were obtained in a mixture of hydrogen and deuterium (n_h/n_d = 0.2) at a frequency slightly below the hydrogen cyclotron frequency. Even though the wave frequency is above the deuterium cyclotron frequency, the torsional Alfven wave can still propagate, due to the hydrogen content, as an ion–ion hybrid wave. The properties of the ion–ion hybrid wave are described in Chapter 13. One of its properties, evident from Fig. 6.5, is that it is guided along steady field lines. This wave may therefore contribute to edge heating, especially if it is guided into a cyclotron resonance layer at the plasma edge where it will be cyclotron damped.

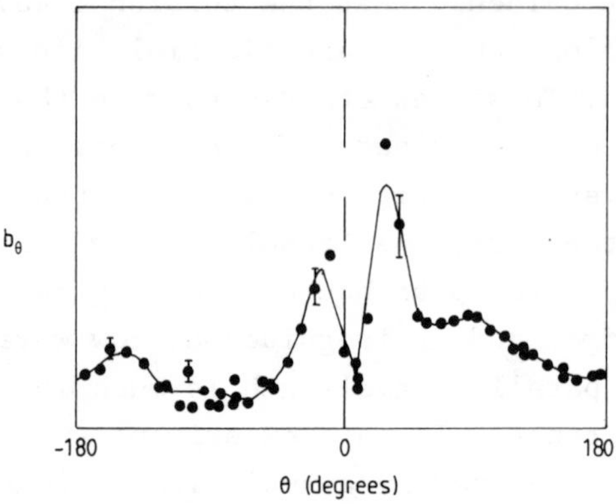

Fig. 6.5 Wave guidance in a plasma containing two ion species, hydrogen (17%) and deuterium (83%), at a frequency near the hydrogen cyclotron frequency.

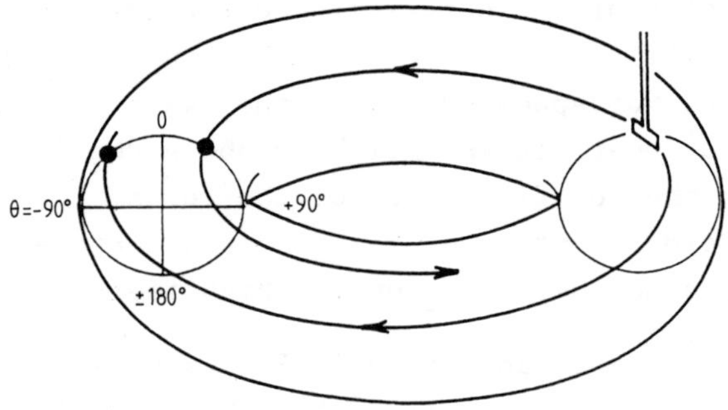

Fig. 6.6 Schematic diagram showing the ray trajectories inferred from the results in Figs. 6.4 and 6.5.

SPACE FILLER

A low frequency torsional Alfven wave does not compress the plasma, but it induces charge separation since $\nabla \cdot \mathbf{E} \neq 0$. Given that a laser beam will scatter off electron density fluctuations, will it scatter off a low frequency torsional wave?

7. REFLECTION AND REFRACTION OF ALFVEN WAVES

The propagation of Alfven waves in a plasma with a density gradient is a rather complex subject but has important consequences for the Alfven wave heating scheme and for ICRF heating. The compressional Alfven wave is refracted by a density gradient in a manner closely analogous to the refraction of a beam of light through a medium of varying refractive index. In effect, the only difference is that the propagation speed is the Alfven speed rather than the speed of light. The torsional Alfven wave is <u>not</u> refracted by a density gradient, at least in the ideal MHD limit, but it may be phase mixed by the gradient. Phase mixing refers to the fact that different sections of a torsional wave front travel at different speeds (the local Alfven speed) along the steady field lines. This process is analysed in Chapter 14 for a smoothly varying density profile.

In this Chapter, we make the simplifying assumption that a smooth density profile can be simulated by a density step or by a series of small density steps. This approach provides a reliable picture of the reflection and refraction of waves by a density gradient, but it removes the possibility of a resonance which can occur in a plasma with a smooth density profile. Resonances will be considered in later Chapters.

Consider the case where **B** is in the z direction, a uniform plasma of ion density n_1 occupies the space $x > 0$ and a uniform plasma of density n_2 occupies the space $x < 0$. We assume that **B** does not vary in the x, y or z directions. The density changes in a direction perpendicular to **B**, corresponding to the usual laboratory situation, and we consider a plane wave incident on the boundary at $x = 0$. Using terminology familiar from optics, the incident, reflected and refracted beams are shown in Fig. 7.1. If $k_y = 0$, all three beams will lie in the x-z plane. If $k_y \neq 0$, all three beams will lie in the same plane, but the normal to this plane is not perpendicular to **B**. We show below that the angle of incidence is equal to the angle of reflection, in the usual optical sense, and a refracted compressional wave obeys

Snell's law. We also obtain the surprising result that there is no refracted torsional wave.

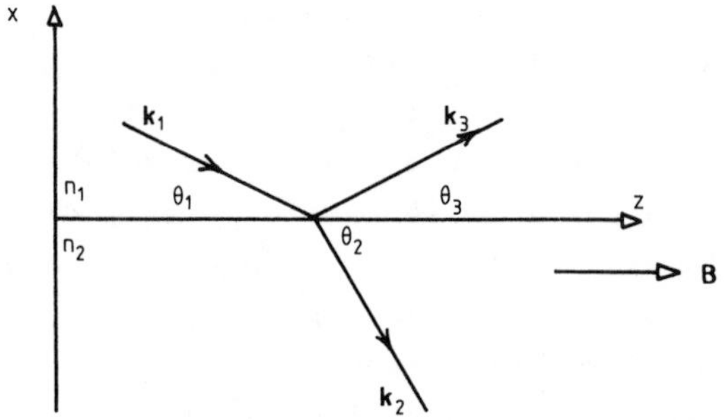

Fig.7.1 Reflection and refraction of Alfven waves at a density step.

The effect of the incident wave is to set up a perturbation in the boundary surface. The perturbation generates reflected and refracted waves having the same frequency as the incident wave and the same wavelengths in directions parallel to the surface. For example, suppose that the wave **E** fields of the incident and reflected beams are given respectively by

$$\mathbf{E}_1 = E_{01}e^{i(\mathbf{k}_1\cdot\mathbf{r} - \omega t)} \quad \text{and} \quad \mathbf{E}_3 = E_{03}e^{i(\mathbf{k}_3\cdot\mathbf{r} - \omega t)}$$

These must be identical functions of **r** at the boundary surface, otherwise it would be impossible to satisfy boundary conditions at all points in the surface. The amplitudes of the two beams can be different, but the wavelengths in the y or z directions must be the same. Hence

$$\mathbf{k}_1 \cdot \mathbf{r} = \mathbf{k}_3 \cdot \mathbf{r} \quad \text{or} \quad (\mathbf{k}_1 - \mathbf{k}_3) \cdot \mathbf{r} = 0$$

Since **r** lies in the surface, the vector $\mathbf{k}_1 - \mathbf{k}_3$ must be normal to the surface, so $\theta_1 = \theta_3$ as shown in Fig. 7.2. At any point in the surface, where x = 0,

$$(\mathbf{k}_1 - \mathbf{k}_3) \cdot \mathbf{r} = (k_{y1} - k_{y3})y + (k_{z1} - k_{z3})z = 0$$

Since this relation holds for all y and z,

$$k_{y1} = k_{y3} \qquad \text{and} \qquad k_{z1} = k_{z3}$$

These conditions can only be satisfied if the angle of incidence equals the angle of reflection and if

$$k_{x3} = -k_{x1} \tag{7.1}$$

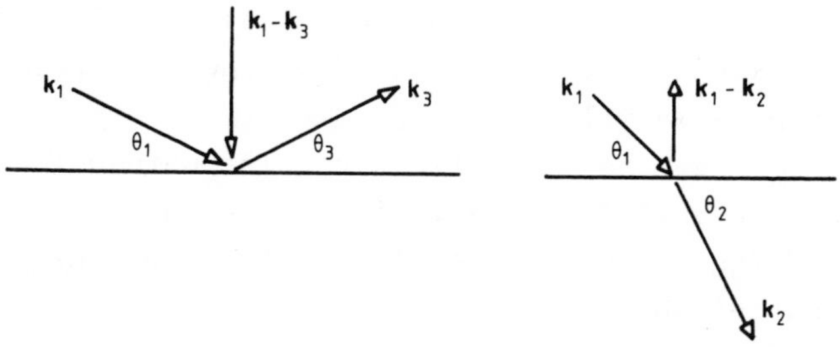

Fig. 7.2 Directions of $\mathbf{k}_1 - \mathbf{k}_3$ and $\mathbf{k}_1 - \mathbf{k}_2$

Similarly, $\mathbf{E}_1$ and $\mathbf{E}_2$ are identical functions of $\mathbf{r}$ in the surface, so $(\mathbf{k}_1 - \mathbf{k}_2)\cdot\mathbf{r} = 0$ and $\mathbf{k}_1 - \mathbf{k}_2$ is normal to the surface as shown in Fig.7.2. Consequently, $\mathbf{k}_1$, $\mathbf{k}_2$ and $\mathbf{k}_3$ all lie in the same plane. At any point in the boundary surface

$$(\mathbf{k}_1 - \mathbf{k}_2) \cdot \mathbf{r} = (k_{y1} - k_{y2})y + (k_{z1} - k_{z2})z = 0$$

$$\therefore \quad k_{y1} = k_{y2} = k_{y3} \qquad \text{and} \qquad k_{z1} = k_{z2} = k_{z3} \tag{7.2}$$

Equation (7.2) is a form of Snell's law. It is only the k_x components which differ either side of the surface. Since $k_z = k \cos\theta$ where θ is the angle between $\mathbf{k}$ and $\mathbf{B}$, Snell's law takes the form

$$\frac{\cos\theta_2}{\cos\theta_1} = \frac{v_2}{v_1}$$

where $v = \omega/k$ is the phase velocity. For the compressional wave, $v = v_A$ so $\cos\theta_2/\cos\theta_1 = v_{A2}/v_{A1} = (n_1/n_2)^{1/2}$. The compressional wave is refracted towards regions of high density and is therefore focussed towards the centre of laboratory plasmas.

For the torsional wave, $v = v_A \cos\theta$ and Snell's law leads to the paradoxical result that $v_{A1} = v_{A2}$. In other words a "refracted" beam can only arise if $n_1 = n_2$. The Poynting vector for this wave is parallel to **B** so there is no energy transport across the density step when **B** is perpendicular to the density step.

The relative amplitudes of the incident, reflected and refracted beams can be calculated by imposing appropriate boundary conditions. At the boundary surface, E_y and b_x are continuous. $E_z = 0$ for Alfven waves in the low frequency MHD approximation. We will also assume that b_z is continuous since there can be no surface current perpendicular to **B** otherwise an infinite $\mathbf{j} \times \mathbf{B}$ force would exist at the surface. Using an obvious notation, these boundary conditions are satisfied if

$$E_{y1} + E_{y3} = E_{y2} \tag{7.3}$$

$$b_{x1} + b_{x3} = b_{x2} \tag{7.4}$$

$$b_{z1} + b_{z3} = b_{z2} \tag{7.5}$$

The b_x, b_z and E_y components are related by (3.6) and (3.11)

$$b_x = -k_z E_y/\omega \tag{7.6}$$

and $$b_z = E_y(k_x^2 + k_y^2)/(\omega k_x) \qquad \text{Compressional} \tag{7.7}$$

or $$b_z = 0 \qquad \text{Torsional}$$

Assuming that the incident, reflected and refracted beams are all compressional waves, (7.1) to (7.7) give:

$$\frac{E_{y3}}{E_{y1}} = \frac{P - Q}{P + Q} \tag{7.8}$$

and $$\frac{E_{y2}}{E_{y1}} = \frac{2P}{P + Q} \tag{7.9}$$

where $P = k_{x2}(k_{x1}^2 + k_y^2)$ and $Q = k_{x1}(k_{x2}^2 + k_y^2)$

If we assume that the incident, reflected and refracted beams are all torsional waves, then the only way to satisfy continuity of k_z is to insist that there is no change of density across the boundary.

An additional possibility is that an incident compressional wave may generate a torsional wave propagating in the region $x < 0$. This is in fact the basis of the Alfven wave heating scheme. However, if $k_y = 0$ there is no mode conversion to the torsional wave since $E_y = b_x = b_z = 0$ for the torsional wave and these are the components which are needed to satisfy the boundary conditions given by Eqs. 7.3 to 7.5. Mode conversion can only occur if $k_y \neq 0$ and if k_z remains continuous across the boundary, ie if $\omega/k_{z1} = v_{A2}$. This process is considered in Chapter 14.

7.1 Ray tracing

Since the torsional wave is not refracted by a density gradient, an isolated ray or a beam of parallel rays simply propagates along the steady field lines. The path of a refracted compressional wave through an inhomogeneous plasma can be estimated by using Snell's law. If v_A varies continuously, the plasma can be regarded as consisting of many thin layers, each of uniform density. Across each layer, $k_z = k\cos\theta$ remains constant where $k = \omega/v_A$, so

$$\cos\theta = (v_A/v_{AO})\cos(\theta_0) \qquad (7.10)$$

where θ_0 is the angle of the incident beam at $x = x_0$ and v_{AO} is the Alfven speed at $x = x_0$. This relation can be used to trace the path of a hypothetical compressional Alfven ray starting at $x = x_0$. The term "ray" here refers to a line drawn in the direction of the group velocity vector. Its trajectory is defined by the relation $d\mathbf{r}/dt = \mathbf{v}_g$. In the MHD limit, the propagation and group velocity vectors are parallel. However, an isolated compressional Alfven ray, analogous to a torsional Alfven ray, cannot be observed in laboratory plasmas since the wavelength of the compressional wave, in a direction perpendicular to **B**, is usually comparable with the plasma dimensions.

Typical ray paths, computed numerically from Eq.(7.10), are shown in Fig. 7.3. The rays turn around, and propagate back into the plasma when $\theta = 0$. For this to occur, the ray must start at a point inside the plasma, rather than at the plasma edge, so that v_A can exceed v_{A0}. Alternatively, the rays can turn around if the steady **B** field increases in the x direction.

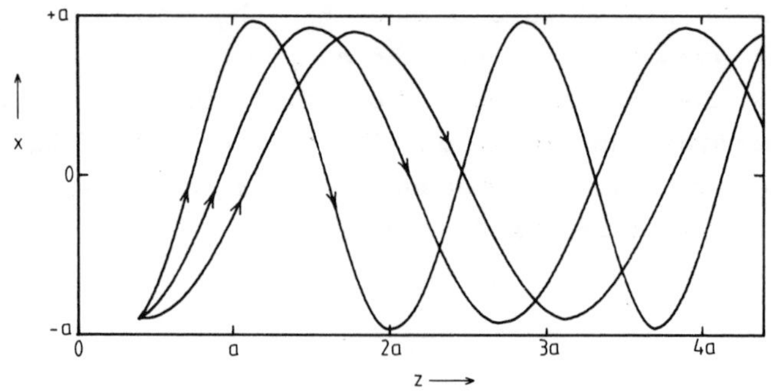

Fig. 7.3 Ray paths for a compressional Alfven wave in a plasma with a parabolic density profile $n = n_0(1 - x^2/a^2)$, for rays starting at $x = -0.9a$. The density is assumed to be independent of y and z.

Ray tracing is generally only useful for qualitative discussion since a significant fraction of the wave energy may be reflected by the density gradient. The ray paths computed from Snell's law refer only to the refracted rays. Strong wave reflections occur if the density varies significantly over a distance equal to a perpendicular wavelength or if there is a surface in the plasma where $k_x = 0$. The latter situation is analysed in Sect. 14.8. Such surfaces exist adjacent to Alfven resonance and ion–ion hybrid resonance surfaces. Alfven ray tracing calculations are strictly only valid in very large devices such as the JET tokamak. In ICRF heating experiments in JET, a significant fraction of the wave energy is absorbed in a single pass of the wave across the plasma. Ray tracing calculations under these conditions provide a very useful picture of the transfer of wave energy from the external antenna.

8. ALFVEN SURFACE WAVES

Consider the situation shown in Fig. 8.1 where a uniform plasma occupies the space $x < 0$, a vacuum exists in the region $x > 0$ and a uniform magnetic field **B** exists in the z direction.

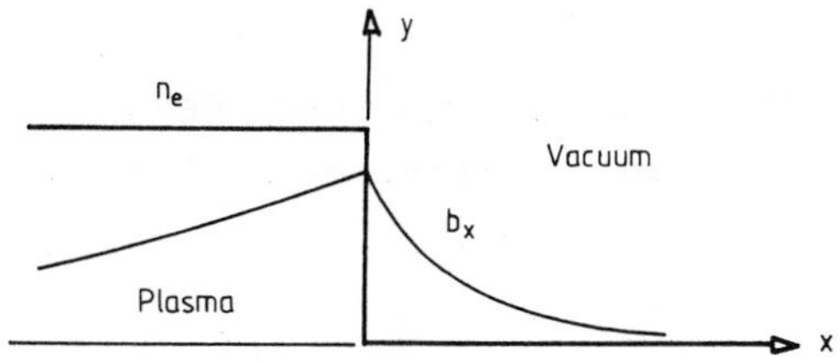

Fig. 8.1 Wave propagation at a plasma–vacuum boundary.

Suppose that a compressional wave is incident on the boundary surface. By analogy with the optical problem of transmission at a glass–vacuum interface, one might expect at least some reflection of the compressional wave. There is no refracted Alfven wave, since a vacuum cannot support an Alfven wave. But neither is there a reflected wave, at least in ideal MHD. The wave fields in a vacuum are normally described with finite displacement current in Maxwell's $\nabla \times \mathbf{b}$ equation. In the case shown in Fig. 8.1, the fields in the vacuum are created by currents in the plasma. The displacement current is negligible at low wave frequencies, both in the plasma and in the vacuum. As a result, the behaviour at a plasma–vacuum boundary is quite different from the behaviour at a glass–vacuum boundary.

For example, if we attempt to apply Snell's law, in the form

$$\frac{\cos\theta_2}{\cos\theta_1} = \frac{v_{A2}}{v_{A1}}$$

then we are in trouble, since $v_{A2} = \infty$ in the vacuum, implying that $\cos\theta_2 = \infty$. Snell's law, in this form, yields a sensible result only if there is a small density jump across the boundary. In optics, the condition $\cos\theta_2 > 1$ corresponds to total internal reflection of the incident beam, the refracted beam being evanescent. However, Snell's law, in the form given by equation (7.2), is always valid. As shown below, application of (7.2) at the plasma–vacuum boundary yields a surface wave solution. The wave fields decrease exponentially on *both* sides of the boundary, as shown in Fig. 8.1. In other words, the wave fields are evanescent not only in the vacuum, but also in the plasma.

Within the plasma, the compressional wave dispersion relation $\omega = kv_A$ can be expressed in the form

$$k_x^2 = \frac{\omega^2}{v_A^2} - k_y^2 - k_z^2 \tag{8.1}$$

It turns out that k_x^2 is negative for the situation shown in Fig. 8.1. Under these conditions, the wave is evanescent in the x direction but propagates in the y and z directions along the surface, with $k_y^2 > 0$ and $k_z^2 > 0$. The wave fields in the plasma are given by equation (3.11);

$$b_{zp} = b_p e^{i(k_x x + k_y y + k_z z - \omega t)} \tag{8.2}$$

$$b_{xp} = -\left(\frac{k_x k_z}{k_\perp^2}\right) b_{zp} \quad \text{where} \quad k_\perp^2 = k_x^2 + k_y^2 \tag{8.3}$$

The vacuum fields can be derived either from Maxwell's equations, ignoring the displacement current, or from the above results by setting $v_A = \infty$. Since k_z and k_y in the vacuum have the same values as those in the plasma (but k_x is different, as described in Chapter 7),

$$\omega^2/k^2 = v_A^2 = \infty \qquad \text{or} \qquad k^2 = k_{xv}^2 + k_y^2 + k_z^2 = 0$$

$$\therefore \quad k_{xv}^2 = -K_v^2 = -(k_y^2 + k_z^2) \tag{8.4}$$

Since $k_\perp^2 = -k_z^2$ when $k^2 = 0$, the vacuum fields are

$$b_{zv} = b_v e^{-K_v x} e^{i(k_y y + k_z z - \omega t)} \tag{8.5}$$

and

$$b_{xv} = \frac{k_{xv}}{k_z} b_{zv} \tag{8.6}$$

where $k_{xv} = +iK_v$ in (8.5) since the vacuum fields must vanish at $x = \infty$. The $\exp(+K_v x)$ solution is therefore excluded. Since b_x and b_z are continuous at the $x = 0$ boundary,

$$b_v = b_p \qquad \text{and} \qquad k_x k_z^2 = -iK_v(k_x^2 + k_y^2) \qquad (8.7)$$

Combining (8.1), (8.4) and (8.7), we obtain

$$(\omega^2/v_A^2 - k_y^2 - k_z^2)k_z^4 = -(k_y^2 + k_z^2)(\omega^2/v_A^2 - k_z^2)^2$$

which simplifies to

$$\frac{\omega^2}{k_z^2} = \left[\frac{2k_y^2 + k_z^2}{k_y^2 + k_z^2}\right] v_A^2 \qquad (8.8)$$

From (8.1) and (8.8), $k_x^2 = -\, k_y^4/(k_y^2 + k_z^2)$, which is obviously negative when k_y^2 and k_z^2 are positive. Equation (8.8) therefore describes a surface wave in which the wave fields decay exponentially either side of the surface. If $k_y \gg k_z$ then $\omega/k_z = \sqrt{2}v_A$. The wave propagates along **B** at $\sqrt{2}$ times the Alfven speed when $k_y \gg k_z$ or at the Alfven speed when $k_y \ll k_z$.

The b_x/b_z ratio in the plasma, from (8.3) and (8.7), is

$$\frac{b_{xp}}{b_{zp}} = \frac{iK_v}{k_z} = i\left[1 + \frac{k_y^2}{k_z^2}\right]^{1/2} \qquad (8.9)$$

For most laboratory conditions, $k_y \gg k_z$ so $b_{zp} \ll b_{xp}$ from (8.9). From equation (3.11), $\nabla\cdot\mathbf{v} = i\omega b_z/B$ so there is only very slight compression of the plasma by the surface wave (WENTZEL, D10). The surface wave therefore has physical properties similar to those of the torsional wave. However, from the above analysis, it is clear that surface waves result from the propagation of a compressional wave along the surface.

In Section 9.2 we will examine surface waves in cylindrical geometry. Unlike conventional waveguide modes, surface waves in a cylindrical plasma do not experience waveguide cutoff and will therefore propagate at frequencies well below the conventional cutoff frequencies. This feature is very important for the Alfven wave heating scheme. The presence of a vacuum is not essential. Surface waves will also propagate if the vacuum is replaced by a low density plasma or if a low density plasma region separates a high density plasma region from a metal wall (D5). The important feature is the surface. The surface itself need not be infinitely steep, since surface waves will propagate at any boundary provided the thickness of the boundary is much smaller than the wavelength in a direction perpendicular to the boundary.

A feature of the Alfven surface wave which has yet to be investigated is the sensitivity of the wave to imperfections in the surface. If it behaves like an acoustic surface wave in a solid it may be extremely sensitive. A strip of adhesive tape placed on a solid surface will damp the surface wave almost completely. The Alfven surface wave is, however, different in one important respect. Real plasmas do not have sharp boundaries. Due to the density gradient, there will generally be a surface within the plasma where the surface wave phase velocity is equal to the local Alfven speed. This leads to Alfven resonance damping of the surface wave, as described in Chapter 14. Imperfections in the surface are therefore unlikely to have such a dramatic effect as that observed in acoustic surface waves.

SPACE FILLER

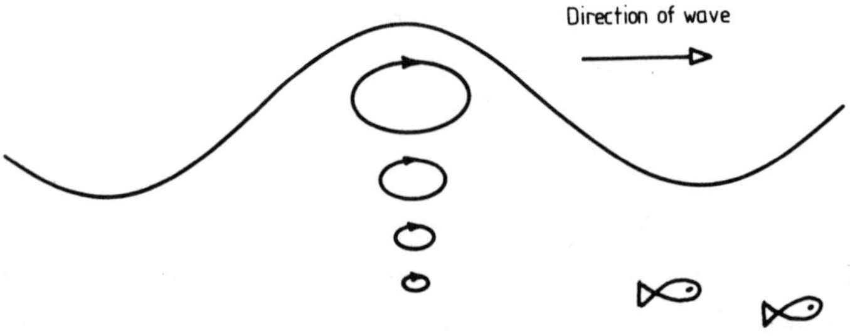

The motion of a particle in a water wave is an ellipse as shown in the diagram. Is this a good picture of an Alfven surface wave?

9. ALFVEN WAVES IN A CYLINDRICAL PLASMA

In the ideal MHD approximation, the behaviour of the torsional Alfven wave is not sensitive to the plasma geometry since this wave propagates only along **B** lines and does not reflect off the plasma boundaries. However, the compressional wave *does* reflect off the plasma boundary or off metal wall boundaries. The geometry is therefore very significant in determining the modes of propagation of the compressional wave. Nevertheless, wave propagation in a toroidal plasma is often modelled in cylindrical geometry, firstly because the mathematics is much simpler and secondly because a torus behaves, in most respects, like a bent cylinder. A good analogy is a bend in a waveguide. Microwaves in a bent waveguide still propagate around the bend due to reflections off the side walls. Provided the bend is not too sharp, the wave fields and phase velocity in the bend are close to the values in a straight guide of the same cross section. In the case of a plasma filled waveguide, turbulence at the plasma boundary and port holes cut into the vessel walls may have significant effects on waveguide modes, but we will assume that the boundaries are perfectly smooth in the following discussion.

In cylindrical geometry we seek solutions of Eqs.(3.1) and (3.2) of the form

$$f(r)e^{i(m\theta + k_{\parallel}z - \omega t)} \tag{9.1}$$

where $k_{\parallel} = k_z$ is the axial wavenumber and m (a positive or negative integer or zero) is the azimuthal mode number, equal to the number of wavelengths in a circumference as shown in Fig. 9.1. The m = 0 mode is azimuthally symmetric.

Fig. 9.1 shows wave magnetic field lines in the plasma cross-section for m = 0, 1 and 2 modes. Also shown are the directions of the unit vectors $\hat{\mathbf{r}}$, $\hat{\theta}$ and $\hat{\mathbf{z}}$. It is conventional to take $\hat{\mathbf{z}}$ to be in the same direction as the steady field, **B**, and to define the positive θ direction by $\hat{\mathbf{r}} \times \hat{\theta} = \hat{\mathbf{z}}$.

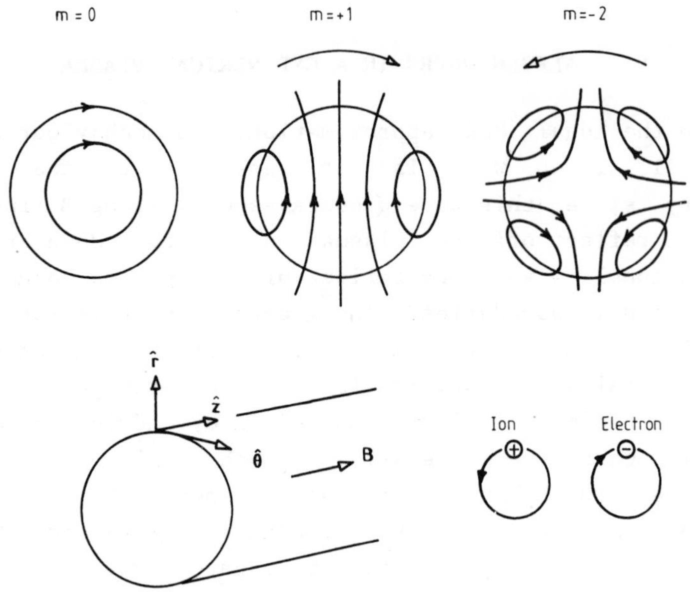

Fig. 9.1 Cylindrical waveguide modes.

The sign of m determines the direction of rotation of the wave fields. Consider a fixed value of z, say z = 0, and assume that **b** varies as $\cos(m\theta - \omega t)$. Then the field maximum rotates in time and is located at $\theta = \omega t/m$. Consequently, $m > 0$ modes rotate in the positive θ direction (in the same sense as the electrons) and $m < 0$ modes rotate in the same sense as positive ions. Some authors use $\exp i(m\theta + k_{\parallel}z + \omega t)$ notation in which case the sense of mode rotation is reversed. An $m = -1$ mode in this notation is then equivalent to an $m = +1$ mode in the $-\omega t$ notation of Eq.(9.1).

The wave field polarisation depends not only on the sign of m but also on the position in the plasma. This is illustrated in Fig. 9.2 for an $m = +1$ mode. The sign of m determines the direction of rotation in a *global* sense but the *local* direction of rotation of the wave fields (i.e. the polarisation) varies with radius and reverses sign at the radius where $b_\theta(r)$ reverses sign. A linearly polarised wave may also propagate in a cylinder. It can be regarded as the sum of an $m = +1$ and an $m = -1$ mode, each having the same amplitude and the same ω and $k_{\parallel}$.

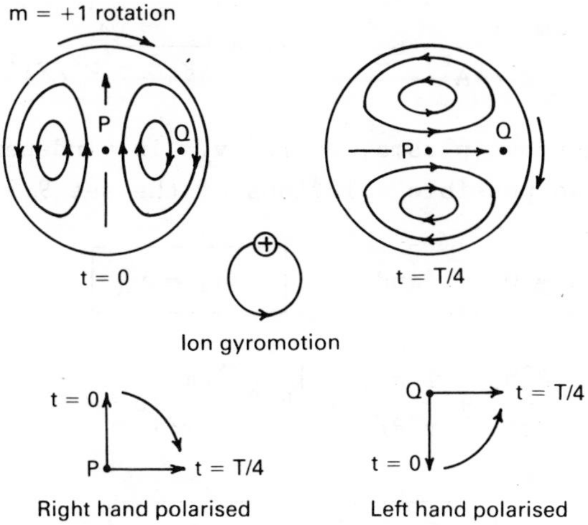

Fig. 9.2 Schematic diagram showing how the global sense of rotation of an m = +1 mode influences the local sense of polarisation. Note that b_r and b_θ are 90^0 out of phase, with respect to time and θ.

9.1 Low frequency dispersion relations

In cylindrical coordinates, the components of Eq.(3.1) are

$$\left.\begin{aligned} -i\omega\rho v_r &= j_\theta B \qquad & E_r &= -v_\theta B \\ i\omega\rho v_\theta &= j_r B \qquad & E_\theta &= v_r B \\ v_z &= 0 \qquad & E_z &= 0 \end{aligned}\right\} \tag{9.2}$$

$$\therefore \qquad i\omega\mu_o j_r = \omega^2 E_r/v_A{}^2 \qquad i\omega\mu_o j_\theta = \omega^2 E_\theta/v_A{}^2 \tag{9.3}$$

Since $E_z = 0$, the components of Eq.(3.2) are

$$\left.\begin{aligned} \mu_o j_r &= i(mb_z/r - k_\| b_\theta) \qquad & \omega b_r &= -k_\| E_\theta \\ \mu_o j_\theta &= ik_\| b_r - \partial b_z/\partial r \qquad & \omega b_\theta &= k_\| E_r \\ \mu_o j_z &= \frac{1}{r}\frac{\partial(rb_\theta)}{\partial r} - \frac{imb_r}{r} \qquad & i\omega b_z &= \frac{1}{r}\frac{\partial(rE_\theta)}{\partial r} - \frac{imE_r}{r} \end{aligned}\right\} \tag{9.4}$$

From (9.3) and (9.4) we find that

$$E_r = \frac{\omega m}{r(k_\parallel{}^2 - \omega^2/v_A{}^2)} b_z \qquad E_\theta = \frac{i\omega}{(k_\parallel{}^2 - \omega^2/v_A{}^2)} \frac{\partial b_z}{\partial r} \tag{9.5}$$

For a homogeneous plasma, where v_A is independent of r, there are two possible solutions of the set 9.3, 9.4, 9.5;

Either $\quad b_z = 0 \quad$ and $\quad \boxed{\omega/k_\parallel = v_A}$ (9.6)

or

$$\frac{\partial^2 b_z}{\partial r^2} + \frac{1}{r}\frac{\partial b_z}{\partial r} + \left[k_\perp{}^2 - \frac{m^2}{r^2}\right] b_z = 0 \tag{9.7}$$

where

$$\boxed{k_\perp{}^2 = \omega^2/v_A{}^2 - k_\parallel{}^2} \tag{9.8}$$

Equation (9.6) describes the torsional wave whose fields can be arbitrary functions of r (provided that $E_\theta = 0$ at a conducting wall). This result is consistent with the fact that there is no energy propagation perpendicular to **B**.

Equations (9.7) and (9.8) describe the compressional wave. The dispersion relation, Eq.(9.8), is identical to Eq.(3.10) obtained in x,y,z coordinates. The dispersion relation is therefore a property of the plasma, not the geometry. The effect of the geometry is to impose a restriction on the allowed values of $k_\perp$. There is no restriction on the allowed values of $k_\parallel$ if the plasma is infinitely long in the z direction. Eq. (9.7) is Bessel's equation. Provided that $k_\perp{}^2 > 0$, the general solution is (see Sect. 16.1)

$$b_z = b_0 J_m(k_\perp r) + b_1 Y_m(k_\perp r) \tag{9.9}$$

where b_0 and b_1 are arbitrary constants. Since $Y_m(k_\perp r) \to \infty$ as $r \to 0$, the wave fields in the plasma remain finite only if $b_1 = 0$. For example, the m = 0 wave fields are given, from Eqs.(9.4) and (9.5), by

$$\left.\begin{aligned} b_z &= b_0 J_0(k_\perp r) \qquad & b_r &= b_2 J_1(k_\perp r) \\ E_\theta &= E_0 J_1(k_\perp r) \qquad & E_r &= E_z = 0 \end{aligned}\right\} \tag{9.10}$$

If the plasma extends to a perfectly conducting wall at $r = d$ then $E_\theta = 0$ at $r = d$ and hence

$$k_\perp d = 3.83,\ 7.02,\ 10.17,\ 13.32,\ \ldots \tag{9.11}$$

since $J_1(x) = 0$ at $x = 3.83$, 7.02, etc. The solution with $k_\perp d = 3.83$ is called the first radial mode. The other solutions represent higher radial modes, each with successively more nodes between $r = 0$ and the wall. The wave field components b_z and b_r for the first radial mode are shown in Fig. 9.3. The field lines are sketched in Fig. 9.7.

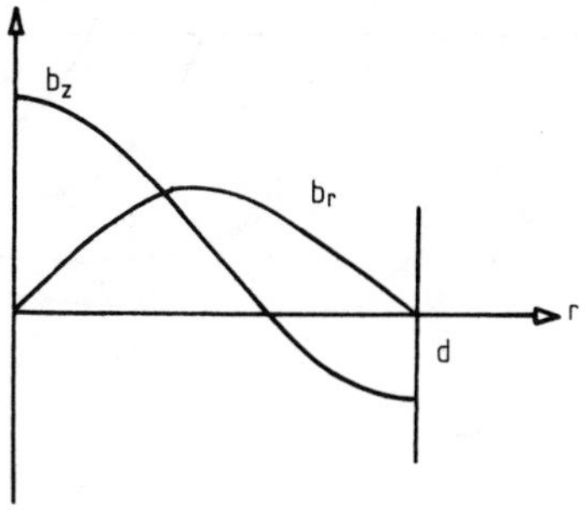

Fig. 9.3 b_z and b_r vs r for the first radial mode $m = 0$ compressional wave in a homogeneous plasma, with a conducting wall at $r = d$.

The various radial modes are waveguide modes which are standing in the r and θ directions but which propagate in the z direction with a phase velocity, from (9.8), given by

$$v_\parallel = \frac{\omega}{k_\parallel} = \frac{v_A}{[1 - (k_\perp v_A/\omega)^2]^{1/2}} \tag{9.12}$$

When $m = 0$, $k_\perp = 3.83/d$ for the first radial mode. When $m \neq 0$, the allowed values of $k_\perp$ must satisfy the condition $db_z/dr = 0$ at $r = d$ (from Eq. (9.5)) and hence

$$\frac{k_\perp d}{m} = \frac{J_m(k_\perp d)}{J_{m-1}(k_\perp d)} \tag{9.13}$$

where
$$\frac{\partial}{\partial r} J_m(k_\perp r) = k_\perp J_{m-1}(k_\perp r) - \frac{m}{r} J_m(k_\perp r) \tag{9.14}$$

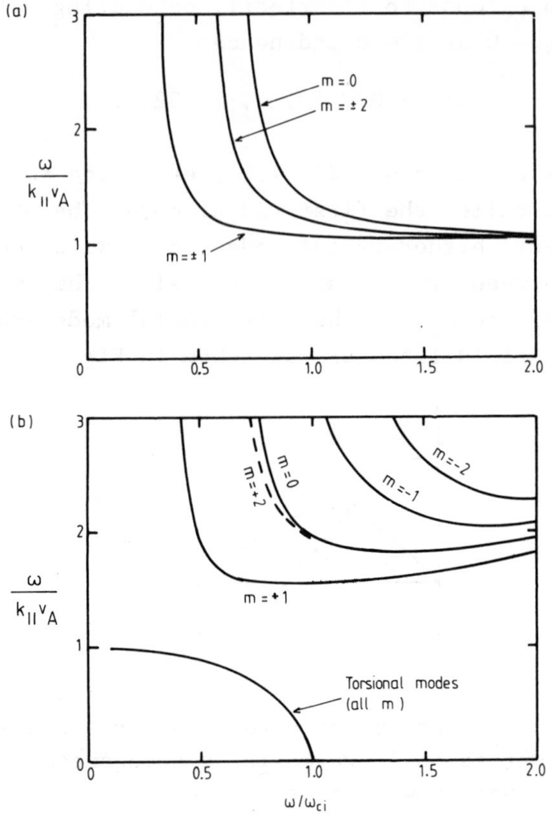

Fig. 9.4 Dispersion relations for a cylindrical plasma with d = 0.2m, n_e = 4 x 10^{19} m^{-3} and where $k_\perp$ is given by (a) the ideal MHD formula, Eq.(9.8) (b) the finite frequency formula, Eq.(10.7) In both cases, $E_\theta(d) = 0$. These solutions are independent of B when plotted in dimensionless form.

Typical dispersion relations for the first radial m = 0, m = ±1 and m = ±2 modes are shown in Fig. 9.4. Each mode has a waveguide cutoff ($k_\parallel = 0$) at $\omega = k_\perp v_A$. At the cutoff frequency the compressional wave propagates purely perpendicular to B with phase velocity $\omega/k_\perp = v_A$. The wave is evanescent, with $k_\parallel^2 < 0$ at frequencies below the cutoff frequency. The solutions given in Fig. 9.4 show that the cutoff frequencies, for typical tokamak conditions, lie near the ion cyclotron frequency. However, the cutoff frequency decreases as the dimension d increases or as the plasma density increases. Consequently, in large tokamak devices, there are many waveguide modes which can propagate at frequencies near or below the ion cyclotron frequency. In the presence of heavy wave damping these modes are not observed experimentally. Under light damping conditions, the modes propagate freely and are observed as cavity eigenmodes when an integer number of wavelengths parallel to B is equal to the circumference of the torus.

We conclude this section with a comment on the wavenumber, $k_\perp$. In slab geometry, $k_\perp^2 = k_x^2 + k_y^2$. In cylindrical geometry, $k_\perp$ cannot be decomposed into r and θ components. One can make a rough attempt by defining $k_\perp^2 = k_r^2 + k_\theta^2$ where $k_\theta = m/r$ as described in Chapter 2. At large r, (9.7) has an approximate solution $b_z \sim \exp(ik_r r)$. This approximation is not very useful, especially when $r \to 0$, since then $k_\theta \to \infty$. The correct variation of b_z with r is given by Eq.(9.9), valid for all values of r.

9.2 Surface waves in cylindrical plasmas

In the previous section, the dispersion relations for Alfven waves were obtained for a plasma with uniform density between r = 0 and the conducting wall. Under most laboratory conditions, the density profile is close to parabolic, dropping to zero at the conducting wall. One can model this profile by assuming that the density is uniform between r = 0 and r = a and that there is a vacuum between the plasma edge (at r = a) and the conducting wall at r = d. This is a rather drastic simplification, but it highlights the fact that new compressional wave modes can propagate, without undergoing cutoff, at frequencies down to zero. These are the surface wave modes examined previously in Chapter 8 for slab geometry. The presence of the wall is not as significant as the presence of the density step. The analysis is simplified if we remove the wall altogether and consider a uniform density plasma between r = 0 and r = a, surrounded by a vacuum extending to $r = \infty$.

Within the plasma, the compressional wave magnetic field components are given by (9.4), (9.5), (9.8) and (9.9),

$$b_z = b_p J_m(k_\perp r) \tag{9.15}$$

$$b_r = \left[\frac{ik_\parallel}{k_\perp^2}\right]\frac{\partial b_z}{\partial r} \tag{9.16}$$

$$b_\theta = -\left[\frac{mk_\parallel}{rk_\perp^2}\right] b_z \tag{9.17}$$

The magnetic field components in the vacuum can be found, as in Chapter 8, by taking $v_A = \infty$ in (9.8) and hence $k_\perp^2 = -k_\parallel^2$ in (9.7). The general solution of (9.7) in this case is

$$b_z = b_1 K_m(k_\parallel r) + b_2 I_m(k_\parallel r) \tag{9.18}$$

where K_m and I_m are modified Bessel functions (see Sect. 16.1). If a wall is near the plasma edge, both functions are required for a solution. The I_m function $\to \infty$ as $r \to \infty$, hence we set $b_2 = 0$ when the plasma is surrounded by a vacuum extending to infinity. The vacuum fields are

$$b_z = b_v K_m(k_\parallel r) \tag{9.19}$$

$$b_r = \frac{1}{ik_\parallel}\frac{\partial b_z}{\partial r} \quad \text{and} \quad b_\theta = \frac{mb_z}{k_\parallel r} \tag{9.20}$$

where $$\frac{\partial K_m(k_\parallel r)}{\partial r} = -k_\parallel K_{m-1}(k_\parallel r) - \frac{m}{r} K_m(k_\parallel r) .$$

Since b_z and b_r are continuous at the boundary we find that

$$b_v = b_p J_m(k_\perp a)/K_m(k_\parallel a)$$

and $$\frac{1}{x^2}\left[\frac{xJ_{m-1}(x)}{J_m(x)} - m\right] = \frac{1}{y^2}\left[\frac{yK_{m-1}(y)}{K_m(y)} + m\right] \tag{9.21}$$

where $x = k_\perp a$ and $y = k_\parallel a$. At low wave frequencies, a simple solution of (9.21) can be obtained assuming that the wavelengths $\lambda_\parallel$ and $\lambda_\perp$ are both much larger than a, so that $x \ll 1$ and $y \ll 1$. The Bessel functions can be expressed in infinite series form, leading to the results that

$$\frac{xJ_{m-1}(x)}{J_m(x)} \to 2m \qquad \text{as } x \to 0$$

and

$$\frac{yK_{m-1}(y)}{K_m(y)} \to 0 \qquad \text{as } y \to 0 .$$

From (9.21) we find that $k_\perp{}^2 = k_\parallel{}^2$ and hence from (9.8) the surface wave dispersion relation is simply

$$\boxed{\omega/k_\parallel = \sqrt{2}\, v_A} \tag{9.22}$$

This agrees with the result found in Chapter 8 for surface waves in slab geometry. The assumption $x \ll 1$ and $y \ll 1$ is seen to be valid when ω is sufficiently small.

There is no surface wave in a cylindrical plasma when $m = 0$ since there is then no solution of Eq. (9.21) for real x and real y. The physical reason is that, for a surface wave mode, current flowing along the surface parallel to **B** must return through another part of the surface (the diametrically opposite part if $m = 1$) but this cannot occur in the case of the azimuthally symmetric $m = 0$ wave. In the case of a "body" or waveguide $m = 0$ mode, current parallel to **B** returns symmetrically through the body of the plasma, not through the surface.

The magnetic field components of the surface wave, with $k_\perp = k_\parallel$, are given by Eqs. (9.15) to (9.17). For example, when $m = 1$ and $k_\perp a \ll 1$, the components are

$$\left.\begin{aligned} b_z &= b_p J_1(k_\perp r) \\ b_\theta &= -\frac{b_z}{k_\perp r} \qquad \text{and} \qquad ib_r = b_\theta \end{aligned}\right\} \tag{9.23}$$

The b_z component is negligible when $k_\perp r \ll 1$ while the b_r and b_θ components are equal in amplitude, 90° out of phase and independent of r (since $J_1(x)/x = 0.5$ when $x \ll 1$). The magnetic field lines therefore appear as shown in Fig.9.5 and the field rotates in the same direction as the electrons if $m > 0$ or in the same direction as the ions if $m < 0$.

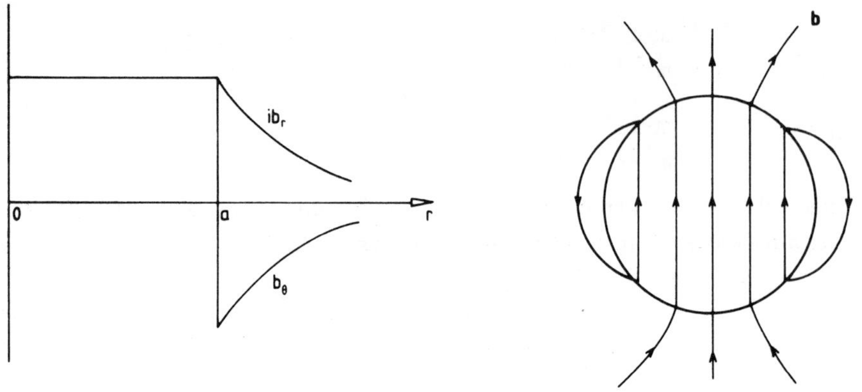

Fig. 9.5 Magnetic field profiles, and two dimensional field plot for an m = 1 surface wave mode at low wave frequencies.

At first sight, it appears that the $m = 1$ wave fields in Fig. 9.5 are not representative of a surface wave. In slab geometry, the wave fields decay exponentially either side of the surface. The same is true in a cylinder, except that the decay length is much longer than the plasma diameter at low wave frequencies. Furthermore, the wave fields from currents on opposite sides of the surface interfere constructively. The latter effect can be visualised more clearly if one considers a rectangular slab of plasma, of finite width, with surface waves on opposing surfaces. If the decay length is much larger than the width of the slab, the wave fields will interfere to produce an essentially uniform field in the plasma.

A correspondence between surface waves in slab and cylindrical geometry can be obtained by defining radial and azimuthal wavenumbers, k_r and $k_\theta = m/r$, such that $k_\perp^2 = k_r^2 + m^2/r^2$ in a cylinder or $k_\perp^2 = k_x^2 + k_y^2$ in a slab. In a WKB approximation, $\partial/\partial r \sim ik_r$. A surface wave is one with $k_x^2 < 0$ in a slab or with $k_r^2 < 0$ in a cylinder. From the above results, $k_r^2 r^2 = k_\perp^2 r^2 - m^2 < 0$ when $k_\perp a \ll 1$.

Another surprising feature of the surface wave is that it has a negligible b_z component, despite the fact that the surface wave is due to propagation of the compressional wave along the plasma–vacuum boundary.

A different result is obtained if the plasma is in contact with a *metal* wall, rather than being surrounded by a vacuum. In that case an incident compressional wave is reflected by the wall, 180° out of phase with respect to the incident wave, to give zero tangential E at the wall. The b_z component is enhanced on reflection as shown in Figs. 9.3 and 9.6. Note that field lines in a surface wave extend into the vacuum, but are confined by a metal wall since they cannot penetrate more than a few skin depths through the wall. The field lines are therefore squashed between the plasma and the wall, enhancing the b_z and b_θ components in this region.

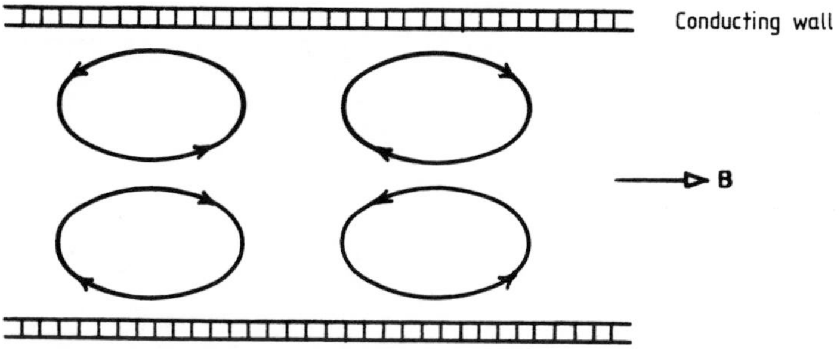

Fig. 9.6 Magnetic field lines for the m = 0 mode in a waveguide. Radial profiles are given in Fig. 9.3.

It is interesting to consider the more general case where a vacuum region separates the plasma from a metal wall. The analysis is straightforward but a bit cumbersome when described in terms of Bessel functions. It is simpler to integrate Eq. (9.7) numerically, using continuity of b_r and b_z to restart the integration at the boundary. This is easily done on a personal computer, as described in Chapter 16. The result for the surface wave is as expected. The presence of the wall enhances the b_z and b_θ components which increase, in relation to b_r, as the wall is brought closer to the plasma edge. This is clearly significant when designing antennas for the Alfven wave heating scheme. An antenna with b_z as its main component would not couple efficiently to a surface wave if the wave had no b_z component. The influence of the wall on the magnetic field components and phase velocity is shown in Fig. 9.7.

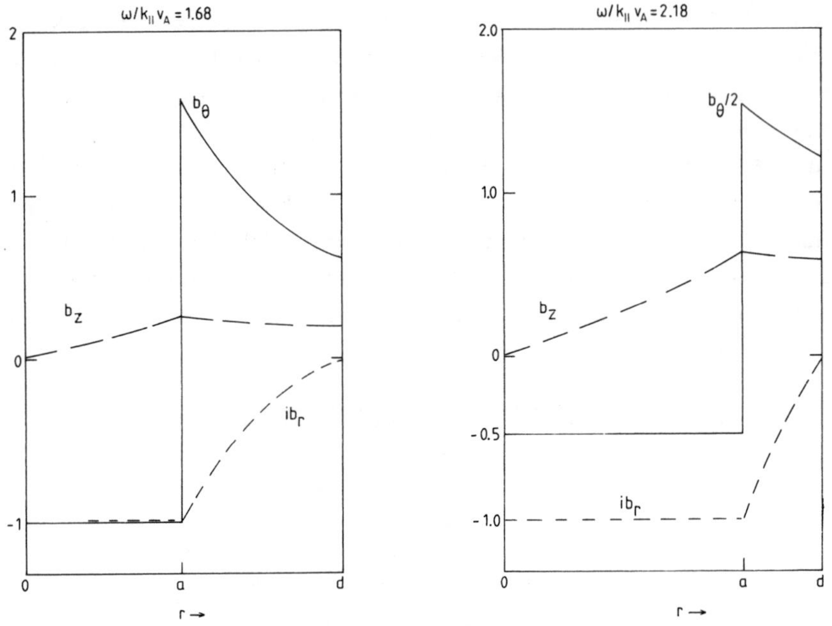

Fig. 9.7 Surface wave fields for an $m = +1$ mode in a hydrogen plasma with $n_e = 4 \times 10^{19}\ m^{-3}$ ($0 < r < a$), a vacuum between $r = a$ and $r = d$, and a conducting wall at $r = d = 0.2$m. In both cases $\omega = 0.1\omega_{ci}$. In the first case, $a =$ 0.10m and $\omega/k_\parallel v_A = 1.68$. In the second case $a = 0.15$m and the phase velocity increases to $\omega/k_\parallel v_A = 2.18$.

SPACE FILLERS

(a) If $b_r = [J_1(k_\perp r)/(k_\perp r)]\sin\theta$, does b_r change sign either side of $r = 0$? The hard part is to decide if $\sin\theta$ should change sign on opposite sides of $r = 0$.

(b) Suppose that an antenna launches an $m = +1$ wave in the z direction. Is the wave launched in the $-z$ direction an $m = +1$ wave or an $m = -1$ wave? If the direction of B is reversed, does the $m = +1$ wave change into an $m = -1$ wave? Interesting results are presented in Refs. C7,F1.

10. ALFVEN WAVES NEAR THE ION CYCLOTRON FREQUENCY

In order to examine the properties of Alfven waves at frequencies near the ion cyclotron frequency, the ideal MHD version of Ohm's law needs to be modified to include the Hall term. At high $k_\perp$ or at frequencies considerably higher than the ion cyclotron frequency, finite electron mass is also important. The general form of Ohm's law, including both effects, as well as plasma resistivity, is given by (SPITZER, A7)

$$\frac{m_e}{n_e e^2}\frac{\partial j}{\partial t} = \mathbf{E} + \mathbf{v} \times \mathbf{B} - \frac{1}{en_e}\,\mathbf{j} \times \mathbf{B} - \eta\mathbf{j} \qquad (10.1)$$

where n_e is the electron density and η is the resistivity.. The analysis of the previous chapters can be extended to include some or all of these extra terms. In this section we consider only those effects introduced by the Hall term, in which case

$$\mathbf{E} + \mathbf{v} \times \mathbf{B} = \frac{1}{en_e}\,\mathbf{j} \times \mathbf{B} \qquad (10.2)$$

The term on the right hand side is called the Hall term since it is responsible for the transverse electric field generated by the Hall effect, first observed in metals and semiconductors. In cylindrical geometry, the components of Eq.(10.2) are

$$\left.\begin{aligned} E_r + v_\theta B &= j_\theta B \,/\, (n_e e) \\ E_\theta - v_r B &= -j_r B \,/\, (n_e e) \\ E_z &= 0 \end{aligned}\right\} \qquad (10.3)$$

Elimination of $\mathbf{v}$ using the left hand set in Eq.(9.2) yields

$$\left.\begin{aligned} i\omega\mu_o j_r &= A\ (\ E_r + i\Omega_i E_\theta\ ) \\ i\omega\mu_o j_\theta &= A\ (\ E_\theta - i\Omega_i E_r\ ) \end{aligned}\right\} \qquad (10.4)$$

where $\Omega_i = \omega/\omega_{ci}$ $\qquad \omega_{ci} = Bq/m_i$ $\qquad q = Ze$

and $A = \dfrac{\omega^2}{v_A{}^2(1 - \Omega_i{}^2)}$ $\qquad v_A{}^2 = \dfrac{B^2}{\mu_o n_i m_i}$ $\qquad n_e = Zn_i$

Eliminating **E** using equation (9.4), we find

$$b_r = \frac{ik_{\parallel}}{k_{\perp}^2}\left(\frac{mG}{rF}b_z + \frac{\partial b_z}{\partial r}\right) \tag{10.5}$$

$$b_\theta = -\frac{k_{\parallel}}{k_{\perp}^2}\left(\frac{m}{r}b_z + \frac{G}{F}\frac{\partial b_z}{\partial r}\right) \tag{10.6}$$

where $F = A - k_{\parallel}^2$, $G = \Omega_i A$

and

$$\boxed{k_{\perp}^2 = \frac{F^2 - G^2}{F}} \tag{10.7}$$

Substituting (10.5) and (10.6) into the $i\omega b_z$ component of (9.4) yields

$$\frac{\partial^2 b_z}{\partial r^2} + \frac{1}{r}\frac{\partial b_z}{\partial r} + \left[k_{\perp}^2 - \frac{m^2}{r^2}\right] b_z = 0 \tag{10.8}$$

Equation (10.7) is the dispersion relation for the plasma. The same relation can be derived in Cartesian coordinates. Equation (10.8) is identical to equation (9.7), except for the finite frequency corrections to $k_{\perp}$ and therefore has the same solution, given by equation (9.9).

The dispersion relation 10.7 is quadratic in $k_{\parallel}^2$ with solutions

$$k_{\parallel}^2 = A - k_{\perp}^2/2 \pm \left[\ \Omega_i^2 A^2 + k_{\perp}^4/4\ \right]^{\frac{1}{2}} \tag{10.9}$$

These two solutions describe the torsional and compressional Alfven waves. When $\Omega_i < 1$, the + sign describes the torsional wave and the − sign describes the compressional wave. When $\Omega_i > 1$, the torsional wave is evanescent ($k_{\parallel}^2 < 0$) and the + sign describes the compressional wave. Solutions of Eq.(10.9) are shown in Fig. 10.1 for typical tokamak conditions, for several values of $k_{\perp}$. These solutions do not necessarily satisfy boundary conditions. Nevertheless, we see from Fig. (10.1) that the phase velocity, $\omega/k_{\parallel}$, of the torsional wave depends only weakly on $k_{\perp}$ and approaches zero as $\omega \to \omega_{ci}$. The compressional wave has a waveguide cutoff frequency which is proportional to $k_\perp$. However, if a vacuum or low density region exists outside the main body of plasma, the compressional wave can propagate as a surface wave at frequencies down to zero (where $k_\perp^2 \to 0$). When boundary conditions are imposed, one obtains solutions (numerically) of the type shown in Fig. 9.4b. Additional solutions are described in Sect. 10.5.

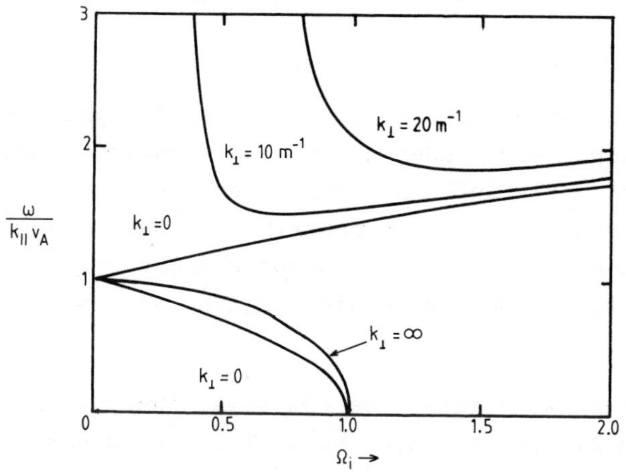

Fig. 10.1 Dispersion relations for torsional and compressional Alfven waves in hydrogen at $n_e = 4 \times 10^{19}\ m^{-3}$. These solutions do not depend on the value of B when plotted in dimensionless units ($\Omega_i = \omega/\omega_{ci}$).

10.1 Propagation parallel or perpendicular to B

If **k** is parallel to **B**, then $k_\perp = 0$, and (10.9) reduces to

$$v_p^2 = \omega^2/k_\parallel^2 = (1 - \Omega_i)v_A^2 \qquad \text{(Torsional)} \qquad (10.10)$$

or

$$v_p^2 = \omega^2/k_\parallel^2 = (1 + \Omega_i)v_A^2 \qquad \text{(Compressional)} \qquad (10.11)$$

The phase velocity is equal to v_A for both wave types when $\Omega_i \ll 1$. The torsional wave undergoes a resonance at the ion cyclotron frequency where $\Omega_i = 1$ and $v_p = 0$. The compressional wave is totally unaffected at $\Omega_i = 1$, since it is then right hand circularly polarised, as described in Sect. 10.3.

If **k** is perpendicular to **B**, then $k_{\parallel} = 0$, and Eq. (10.7) has only one solution, $\omega/k_{\perp} = v_A$, which describes the compressional wave. This solution remains valid at frequencies from zero up to about 10 times the ion cyclotron frequency. At higher frequencies, the finite electron mass term in Ohm's law becomes significant and there is a resonance at the lower hybrid frequency, $\omega \approx (\omega_{ce}\omega_{ci})^{1/2}$.

10.2 The Alfven wave resonance

Equation (10.7) indicates that $k_{\perp}^2 = \infty$ when $F = 0$. Then

$$v_p = \omega/k_{\parallel} = v_A(1 - \Omega_i^2)^{1/2} \tag{10.12}$$

This relation, shown in Fig. 10.1, gives the phase velocity of the torsional wave under the unusual condition where $k_{\parallel}$ is finite but $k_{\perp}$ is infinite. This condition is known as the Alfven wave resonance, also described in the literature as the shear wave resonance, the perpendicular ion cyclotron resonance, the mode conversion surface or the confluence region.

An Alfven wave resonance can arise when v_A varies with radius, usually due to the variation in density. At a particular radius in the plasma where the local Alfven speed is equal to the phase velocity of the compressional mode, energy is extracted from the compressional wave and mode converted to the torsional wave. This is the basis of the Alfven wave heating scheme.

10.3 Wave field components and polarisation

The wave magnetic field components are given by equations (9.9), (10.5) and (10.6) for both Alfven wave types. For example, when $m = 0$,

$$b_z = b_0 J_0(k_{\perp} r)$$

$$b_r = -\frac{ik_{\parallel}}{k_{\perp}} b_0 J_1(k_{\perp} r) \qquad b_\theta = \frac{Gk_{\parallel}}{Fk_{\perp}} b_0 J_1(k_{\perp} r)$$

so
$$\frac{ib_r}{b_\theta} = \frac{F}{G} = \frac{(A - k_{\parallel}^2)}{\Omega_i A} \tag{10.13}$$

In contrast to the MHD behaviour, the torsional wave develops a significant b_z component at finite frequency. The ratios $|b_r/b_\theta|$ and $|b_z/b_\theta|$ are shown in Fig. 10.2 for a hydrogen plasma with $n_e = 4 \times 10^{19} m^{-3}$ taking the peak values $|J_0| = 1.0$ and $|J_1| = 0.58$ since J_0 and J_1 peak at different radii.

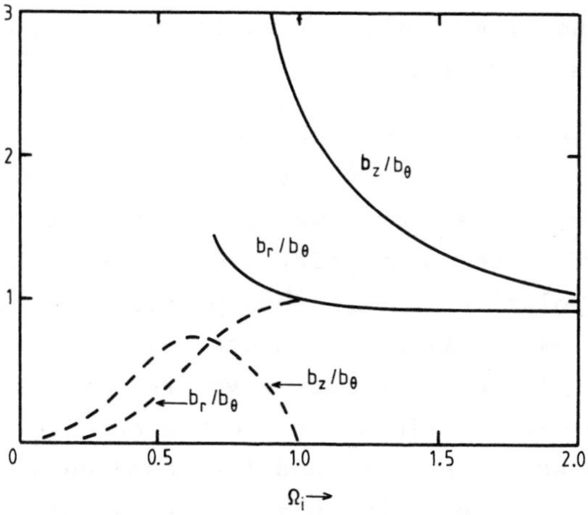

Fig. 10.2 Ratio of magnetic field components for torsional (dashed curves) and compressional (solid curves) m = 0 waves in hydrogen, $n_e = 4 \times 10^{19}$ m^{-3}, d = 0.2m, $k_\perp = 19.15$ m^{-1}.

The behaviour of ib_r/b_θ near the cyclotron frequency is of particular interest. Since $A \rightarrow \infty$ as $\Omega_i \rightarrow 1$, we find from Eq.(10.9) that $k_\parallel^2 \rightarrow 2A$ and $F/A \rightarrow -1$ for the torsional wave. For the compressional wave, $k_\parallel^2 \rightarrow A(1 - \Omega_i) - k_\perp^2/2$, which remains finite as $\Omega_i \rightarrow 1$ so $F/A \rightarrow +1$.

Consequently, $ib_r/b_\theta = -1$ for the torsional wave, and $ib_r/b_\theta = +1$ for the compressional wave, at $\Omega_i = 1$. The torsional wave is left hand circularly polarised, the compressional wave being right hand circularly polarised. At other frequencies, these waves are elliptically polarised. To relate the sense of rotation to that of the ions or the electrons, consider ion gyromotion in the absence of an **E** field, so

$$m \frac{d\mathbf{v}}{dt} = q\mathbf{v} \times \mathbf{B}$$

$$-i\omega m v_r = q v_\theta B \qquad \text{and} \qquad -i\omega m v_\theta = -q v_r B$$

$$\therefore \qquad i v_r / v_\theta = -1 \qquad \text{since} \qquad \omega = \omega_{ci} = qB/m_i$$

The ions rotate in a left hand sense. Consequently, it is the torsional wave which has a resonance at the ion cyclotron frequency. The ions can therefore be heated preferentially by RF heating at the ion cyclotron frequency (or at the cyclotron harmonics).

The above results were obtained for $m = 0$ wave types. The behaviour of $m \neq 0$ waves is more complicated. For example, $m = +1$ torsional waves and $m = +1$ compressional waves are BOTH right hand polarised near $r = 0$ and are BOTH left hand polarised near the plasma edge for the first radial mode, as described in Chapter 9. The reverse is true of the $m = -1$ modes. At low wave frequencies, the region "near" $r = 0$ extends almost to the plasma edge so $m = +1$ and $m = -1$ modes are right hand and left hand polarised respectively over most of the plasma cross section. As the frequency is increased towards the ion cyclotron frequency, the region over which a torsional mode is left hand polarised expands, occupying the entire plasma cross section when $\Omega_i = 1$. Similarly, the region over which a compressional mode is right hand polarised expands, occupying the entire plasma cross section when $\Omega_i = 1$. Alternatively, the region over which the $m = -1$ compressional wave is left hand polarised shrinks as ω increases. The $m = -1$ surface wave fields are entirely excluded from the plasma at a frequency about half the ion cyclotron frequency (D2, D5).

10.4 The group velocity vector

The group velocity vector for Alfven waves in the MHD limit was calculated in Chapter 4. In this section we consider finite frequency corrections in Cartesian rather than in cylindrical coordinates, since the wave modes in a cylinder are standing in the radial direction and the time average energy flux is zero in the r direction. The dispersion relation at finite frequency is given, from (10.7), by

$$Fk_\perp^2 = F^2 - G^2 \qquad\qquad F = A - k_\parallel^2$$

Differentiating first with respect to $k_\perp$ and then with respect to $k_\parallel$, we find

$$\frac{\partial\omega}{\partial k_\perp}\left[k_\perp^2 \frac{\partial F}{\partial\omega} - 2F\frac{\partial F}{\partial\omega} + 2G\frac{\partial G}{\partial\omega}\right] = -2Fk_\perp$$

$$\frac{\partial\omega}{\partial k_\parallel}\left[k_\perp^2 \frac{\partial F}{\partial\omega} - 2F\frac{\partial F}{\partial\omega} + 2G\frac{\partial G}{\partial\omega}\right] = 2k_\parallel(k_\perp^2 - 2F)$$

The ratio, G_v, of the group velocity components is therefore

$$G_v = \frac{\partial\omega/\partial k_\perp}{\partial\omega/\partial k_\parallel} = \frac{Fk_\perp}{k_\parallel(2F - k_\perp^2)} \qquad (10.14)$$

This ratio is plotted in Fig. 10.3 vs Ω_i for both wave types and for two relatively low values of $k_\perp$. The MHD results are only slightly modified at finite frequency. Energy in the torsional wave propagates mainly parallel **B**, especially at high $k_\perp$, and G_v does not exceed about 0.2 even at low $k_\perp$. Energy in the compressional wave propagates approximately parallel to **k**, with $G_v \sim k_\perp/k_\parallel$. Phase and group velocity surfaces for Alfven waves at finite frequency are shown in Fig. 5.3.

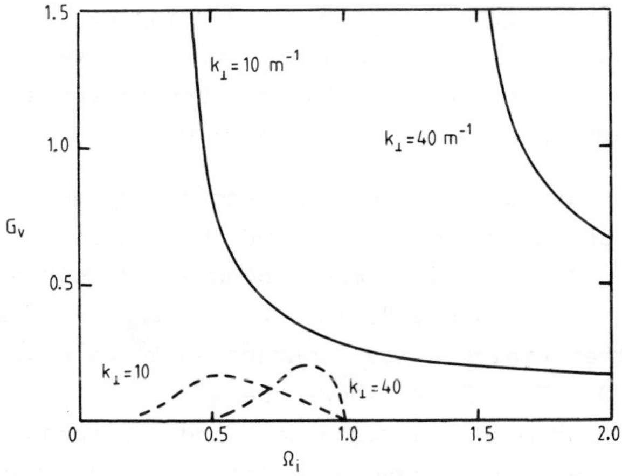

Fig. 10.3 G_v vs Ω_i for torsional (dashed curves) and compressional (solid curves) Alfven waves in hydrogen at $n_e = 4 \times 10^{19}\ m^{-3}$.

These results help to explain why there is no waveguide cutoff for the torsional wave, even though G_V is no longer zero at finite frequency. One might expect torsional waves to reflect off the plasma boundary when G_V is finite and hence experience a waveguide cutoff similar to that experienced by the compressional wave. However, at a waveguide cutoff, $k_\parallel = 0$ and wave energy then propagates entirely perpendicular to **B**. There is no cutoff for torsional or surface wave modes since the wave energy is directed essentially parallel to **B** and since these waves cannot propagate when $k_\parallel = 0$.

10.5 Finite frequency surface waves

It is easy to extend the results of Section 9.2 to include finite frequency effects. Consider a cylindrical plasma of radius a, ion density n_1, surrounded by a concentric perfectly conducting wall of radius d. Assume that the region $a < r < d$ is occupied by plasma of density $n_2 < n_1$. A uniform axial magnetic field $\mathbf{B} = \hat{\mathbf{z}}B$ fills the whole volume. The wave fields in each uniform density region are given by Eqs. 10.5, 10.6 and 10.8. Analytical solutions can therefore be found in terms of Bessel functions, using continuity of b_r and b_z at $r = a$ to match the solutions in each region and using the boundary condition $b_r = 0$ at $r = d$. It is simpler, however, to integrate Eq. (10.8) numerically from $r = 0$ to $r = a$ and then from $r = a$ to $r = d$ using continuity of b_r and b_z to restart the integration. For given values of n_1, n_2, B and ω, one guesses $k_\parallel$ and then corrects the guess until a solution is found with $b_r(d) = 0$.

Typical results for the first radial compressional wave modes are shown in Figs. 10.4 and 10.5 (from Ref. D5). At low wave frequencies, these modes behave as surface waves with $k_r^2 < 0$ (see Sect. 9.2). At higher wave frequencies ($\Omega_i > 0.5$) they evolve into conventional waveguide modes with $k_r^2 > 0$. Fig. 10.4 shows the effect of varying n_2 on the m = +1 and m = −1 surface wave modes respectively. The m = +1 mode, being left-hand polarized near the plasma edge, undergoes a resonance at the ion cyclotron frequency (provided $n_2 \neq 0$) but reappears at frequencies above the cyclotron frequency as a conventional compressional waveguide mode. The m = −1 mode, being right-hand polarised near the plasma edge, is largely unaffected by changes in n_2.

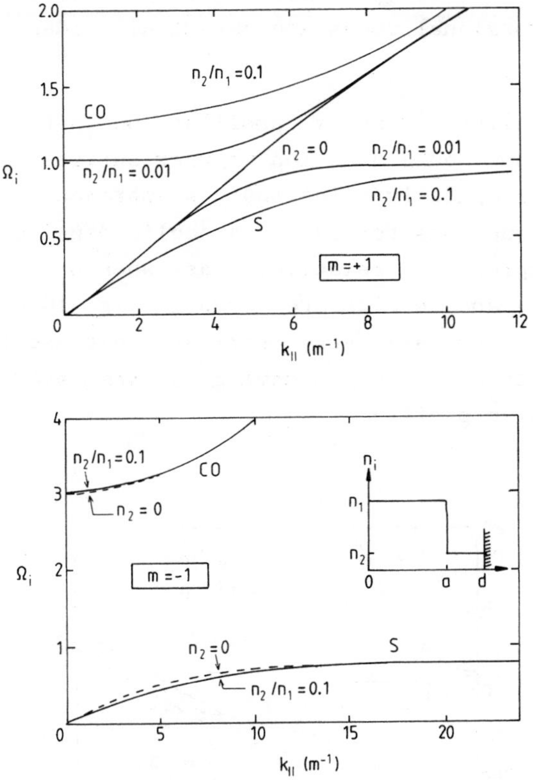

Fig. 10.4. Dispersion relations for the first radial mode m = +1 and m = −1 compressional waves in a step density profile (see inset) with d = 0.12 m, a/d = 0.8, showing cutoff (CO) and surface wave (S) branches, for a hydrogen plasma with $n_1 = 1 \times 10^{19}$ m^{-3}.

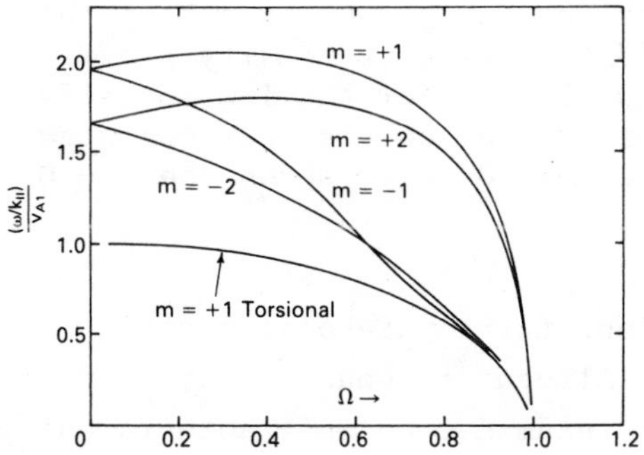

Fig. 10.5 Dispersion relations for the first radial mode m = ± 1, ± 2 compressional waves (same conditions as Fig.10.4) with $n_2/n_1 = 0.1$. Also shown is the m = +1 torsional wave.

Dispersion relations for $m = \pm 1$ and $m = \pm 2$ surface wave modes are compared in Fig. 10.5. Also shown is the first radial mode $m = +1$ torsional wave mode. Dispersion relations for other torsional modes are essentially identical to the $m = +1$ mode.

Since realistic density profiles do not have sharp boundaries, one could question whether surface waves can be observed experimentally. The obvious approach is to reformulate the equations for say a parabolic density distribution. This approach is relatively easy when $\Omega_i > 1$ (as shown in Sect. 14.7) and results are given in Fig. 10.6 (from Ref. D5). The results are very similar to those found for a uniform density distribution having the same average density as the parabolic profile.

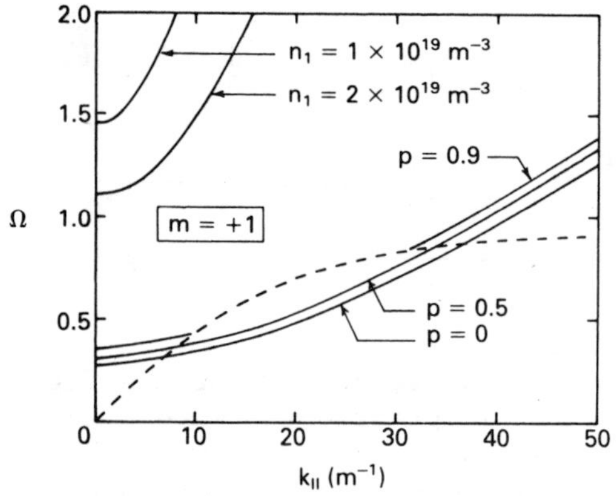

Fig. 10.6 Dispersion relations for the CO branch of the $m = +1$ compressional wave, for a parabolic density profile $n_i = n_1\ (1 - pr^2/d^2)$, $d = 0.12$ m. For the low density curves, $p = 0.98$. The lower three curves are computed with $n_1 = 2 \times 10^{20}\ m^{-3}$. The dashed curve shows the $F = 0$ resonance with $p = 0.9$.

When $\Omega_i < 1$, the equations for a parabolic profile are singular if there is an Alfven resonance layer in the plasma. This problem has been examined by many authors using various theoretical approaches. For example, APPERT et. al.

(H2) find that the dispersion relation can be deduced from calculations of the impedance of an antenna surrounding the plasma. The real part of the impedance is due to Alfven resonance damping of the surface wave (see Section 14.4). The results indicate that, except for wave damping, the qualitative features of surface wave modes in a parabolic profile are the same as those in a step density profile.

Suppose then that one approximates a parabolic profile by a series of density steps. Results for two steps are given in Fig. 10.7 and results for 9 steps are given in Fig. 10.8. It is found that each new step introduces a new wave mode. At first sight, it would appear that each surface is acting to support a surface wave mode. This is clearly the case for a two step profile. However, when one examines the various modes for a large number of steps, it is found that each mode has a field maximum at a different step and it is only the modes localised near the plasma edge which behave as surface wave modes. Modes localised away from the edge of the plasma can be identified as torsional wave modes. This behaviour is consistent with the results for a smooth

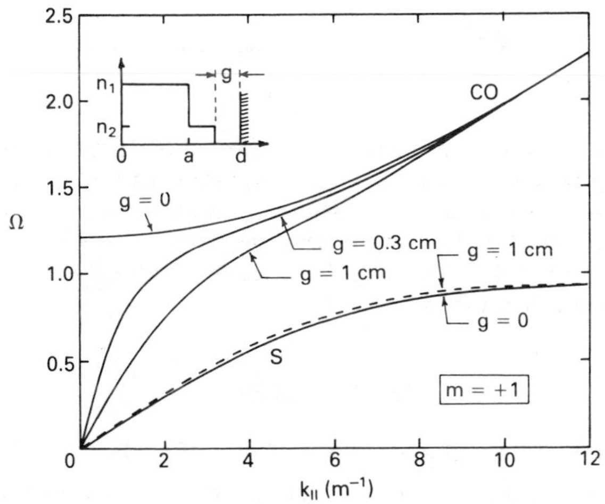

Fig. 10.7 Dispersion relations for the first radial mode m = + 1 compressional wave in a step density profile (see inset) with $n_1 = 1 \times 10^{19}$ m^{-3}, $n_2/n_1 = 0.1$, d = 0.12m, a/d = 0.8, and with vacuum gaps as indicated.

density profile (H2), where it was found that a single surface wave mode was embedded in a continuum of torsional wave modes. Similar behaviour is also found for electrostatic waves in an inhomogeneous plasma, which also has a continuous spectrum, due to the continuous variation of the plasma frequency.

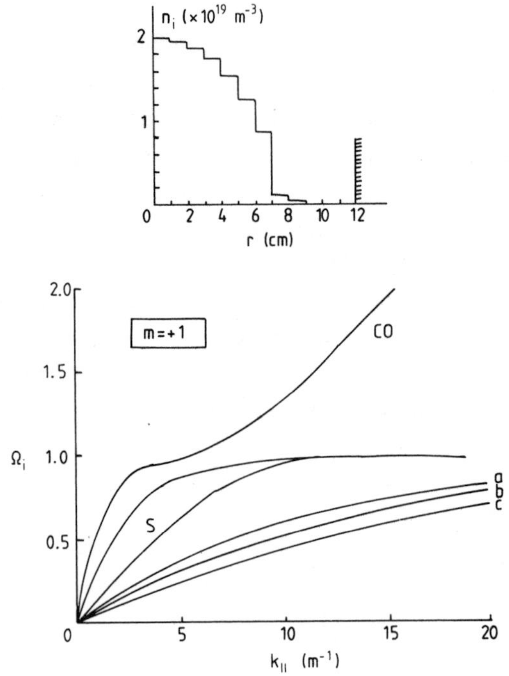

Fig. 10.8 Dispersion relations for the first radial mode m = + 1 eigenmodes in an N = 9 step density profile. Only six of the ten modes are shown. Modes a, b, and c are torsional modes localised in the regions r = 6–7, 5–6, and 0–1 cm respectively. The surface wave, labelled S, evolves into a torsional wave as $\Omega_i \to 1$.

SPACE FILLER

Why does the phase velocity of the torsional wave *decrease*, and the velocity of the compressional wave *increase*, as the wave frequency increases towards the ion cyclotron frequency? Can this be explained in terms of magnetic pressure or tension or effective ion mass?

11. ALFVEN WAVES IN A CURRENT CARRYING PLASMA

In a tokamak plasma, the poloidal field, B_θ, generated by the ohmic heating current is much smaller than the toroidal field, B_z. One might therefore expect that the plasma current would have an insignificant effect on Alfven wave propagation. For example, if $B_\theta/B_z = 0.1$, the total field is only 0.5% larger than B_z, so the Alfven speed is essentially the same as in a current-free plasma. Nevertheless, the plasma current has a surprisingly strong effect, due to the fact that wave energy is guided along helically twisted field lines, as shown in Fig. 11.1.

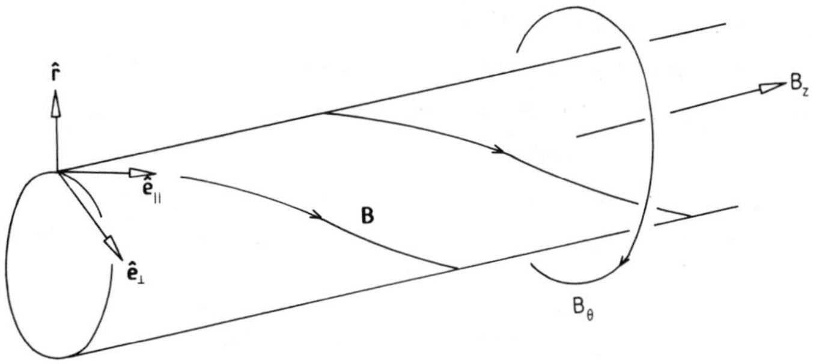

Fig. 11.1 Cylindrical current-carrying plasma.

The effect of the current can be estimated in cylindrical geometry from the equation of motion and Ohm's law

$$\rho \frac{\partial \mathbf{v}}{\partial t} = \mathbf{j} \times \mathbf{B} + \mathbf{J} \times \mathbf{b} \tag{11.1}$$

$$\mathbf{E} + \mathbf{v} \times \mathbf{B} = \frac{1}{n_e e} (\mathbf{j} \times \mathbf{B} + \mathbf{J} \times \mathbf{b}) \tag{11.2}$$

where $\mathbf{B}$ is the steady magnetic field, $\mathbf{J}$ is the zero-order current density and $\mathbf{b}$, $\mathbf{j}$ are small perturbations.

Let
$$\mathbf{B} = \hat{\mathbf{e}}_\parallel B_0 = \hat{\theta} B_\theta + \hat{\mathbf{z}} B_z \tag{11.3}$$

where $B_0{}^2 = B_\theta{}^2 + B_z{}^2$ and $\hat{\mathbf{e}}_\parallel$ is a unit vector locally parallel to $\mathbf{B}$.

In a toroidal, current – carrying plasma, it is conventional to specify the helical twist of the field lines in terms of the safety factor $q(r) = rB_z/[RB_\theta(r)]$ where R is the major radius of the torus. For simplicity, we now assume that J is independent of r, in which case $B_\theta(r) = \mu_0 Jr/2$ and hence q is independent of r. This is a zero shear model since all field lines rotate through the same angle θ in traversing a given distance z, regardless of r. We also assume that B_z and n_e are uniform and that $B_\theta \ll B_z$.

Solutions of the above equations are further simplified if a local cylindrical coordinate system $(r, e_\perp, e_\parallel)$ is used (E3,H2) instead of (r,θ,z). The unit vector $\hat{\mathbf{r}}$ is the same in both coordinate systems, while $\hat{\mathbf{e}}_\perp$ is defined by the relation $\hat{\mathbf{r}} \times \hat{\mathbf{e}}_\perp = \hat{\mathbf{e}}_\parallel$. In the (r,θ,z) coordinate system, one can assume perturbations of the form $f(r)\exp[i(k_z z+m\theta-\omega t)]$ or $f(r)\exp[i(k_z z+k_\theta s-\omega t)]$ where $s = r\theta$ is an arc length and $k_\theta = m/r$ is the azimuthal wave number. This is sensible since $k_\theta s$ is independent of r. In the $(r,e_\perp,e_\parallel)$ coordinate system, small perturbations can be expressed in the form

$$f(r)e^{i(k_\parallel e_\parallel + m_\perp e_\perp - \omega t)} \tag{11.4}$$

To find relations between the various quantities in the two coordinate systems, we take the dot and cross products of **k** and **B**;

$$\mathbf{k} \cdot \mathbf{B} = k_\parallel B_0 = \frac{m}{r} B_\theta + k_z B_z \tag{11.5}$$

$$(\mathbf{k} \times \mathbf{B})_r = \frac{m_\perp B_0}{r} = \frac{m}{r} B_z - k_z B_\theta \tag{11.6}$$

A torus of major radius R can be approximated in cylindrical geometry using the boundary condition that the wave fields must be periodic in a distance $2\pi R$. In this case $n\lambda = 2\pi R$ where n is an integer, so $k_z = 2\pi/\lambda = n/R$. For an infinitely long cylinder without such a periodic boundary condition, n is continuously variable and R is some convenient scale length.

$$\therefore \quad k_{\parallel} = \frac{B_z}{B_o}\left(\frac{n}{R} + \frac{mB_\theta}{rB_z}\right) = \frac{1}{R}\left(n + \frac{m}{q}\right) + O(B_\theta/B_z)^2 \qquad (11.7)$$

When $B_\theta \ll B_z$,

$$m_\perp = \left(1 - \frac{B_\theta{}^2}{2B_z{}^2}\right)\left(m - \frac{nr}{R}\frac{B_\theta}{B_z}\right) = m + O(B_\theta/B_z)^2 \qquad (11.8)$$

where r/R is a quantity of order B_θ/B_z in tokamak devices. Provided that $|n/m| \leq 10$, the variation of $m_\perp$ with r is negligible. Similarly, for a zero shear current profile, the variation of $k_\parallel$ with r is negligible and the exponent in Eq.(11.4) has no significant dependence on r.

Now let $\mathbf{J} = \hat{\mathbf{e}}_\parallel J$ so that the zero order current is parallel to $\mathbf{B}$ and the equilibrium is force-free ($\mathbf{J} \times \mathbf{B} = 0$). Let $\mathbf{b} = (b_r, b_\perp, b_\parallel)$, $\mathbf{j} = (j_r, j_\perp, j_\parallel)$ etc. so that

$$\mathbf{J} \times \mathbf{b} = \hat{\mathbf{r}}(-Jb_\perp) + \hat{\mathbf{e}}_\perp(Jb_r)$$

$$\mathbf{j} \times \mathbf{B} = \hat{\mathbf{r}}(j_\perp B_o) + \hat{\mathbf{e}}_\perp(-j_r B_o)$$

From equations (11.1) and (11.2) we find that

$$E_\parallel = 0$$

$$i\omega\mu_o j_r = AE_r + iGE_\perp \qquad (11.9)$$

$$i\omega\mu_o j_\perp = AE_\perp - iGE_r \qquad (11.10)$$

$$A = \frac{\omega^2}{v_A{}^2(1-\Omega^2)} \qquad \Omega = \frac{\omega}{\omega_{ci}} \qquad v_A{}^2 = \frac{B_o{}^2}{\mu_o \rho}$$

$$G = \Omega A - Dk_\parallel \qquad D = \mu_o J/B_o$$

The components of Maxwell's curl equations in the local cylindrical coordinate system, neglecting the displacement current, are

$$E_r = \frac{\omega b_\perp}{k_\parallel} \qquad E_\perp = -\frac{\omega b_r}{k_\parallel} \tag{11.11}$$

$$i\omega\mu_o j_r = -\frac{\omega m_\perp}{r} b_\parallel + \omega k_\parallel b_\perp \tag{11.12}$$

$$i\omega\mu_o j_\perp = -\omega k_\parallel b_r - i\omega \frac{\partial b_\parallel}{\partial r} \tag{11.13}$$

From (11.9) to (11.13) we find that

$$b_r = \frac{ik_\parallel}{k_\perp{}^2}\left(\frac{m_\perp G}{rF} b_\parallel + \frac{\partial b_\parallel}{\partial r}\right) \tag{11.14}$$

$$b_\perp = -\frac{k_\parallel}{k_\perp{}^2}\left(\frac{m_\perp}{r} b_\parallel + \frac{G}{F}\frac{\partial b_\parallel}{\partial r}\right) \tag{11.15}$$

where $\qquad F = A - k_\parallel{}^2$

and

$$\boxed{k_\perp{}^2 = \frac{F^2 - G^2}{F}} \tag{11.16}$$

Substitution into the z component of $\nabla \times \mathbf{E} = i\omega\mathbf{b}$ gives

$$\frac{\partial^2 b_\parallel}{\partial r^2} + \frac{1}{r}\frac{\partial b_\parallel}{\partial r} + \left[k_\perp{}^2 - \frac{m_\perp{}^2}{r^2}\right] b_\parallel = 0 \tag{11.17}$$

which is consistent with the relation

$$\nabla \cdot \mathbf{b} = \frac{1}{r}\frac{\partial}{\partial r}(rb_r) + \frac{im_\perp b_\perp}{r} + ik_\parallel b_\parallel = 0 \tag{11.18}$$

Equation (11.16) is the dispersion relation for Alfven waves in a zero shear, current-carrying plasma, including finite frequency terms. The variation of $b_\parallel$ with r is given to a good approximation by the solution of Eq. (11.17),

$$b_\parallel = b_o J_{m_\perp}(k_\perp r) \qquad \text{when} \quad k_\perp{}^2 > 0 \tag{11.19}$$

This is not an exact solution since $m_\perp$ varies slightly with r. However, the variation of $m_\perp$ with r is negligible when $B_\theta \ll B_z$.

11.1 Dispersion Relations

Solutions of (11.16) are given in Fig. 11.2 for typical tokamak conditions, with D = 2/(qR) = 1 m^{-1}. These solutions are meant to illustrate the effect of the plasma current but do not necessarily satisfy boundary conditions. The boundary

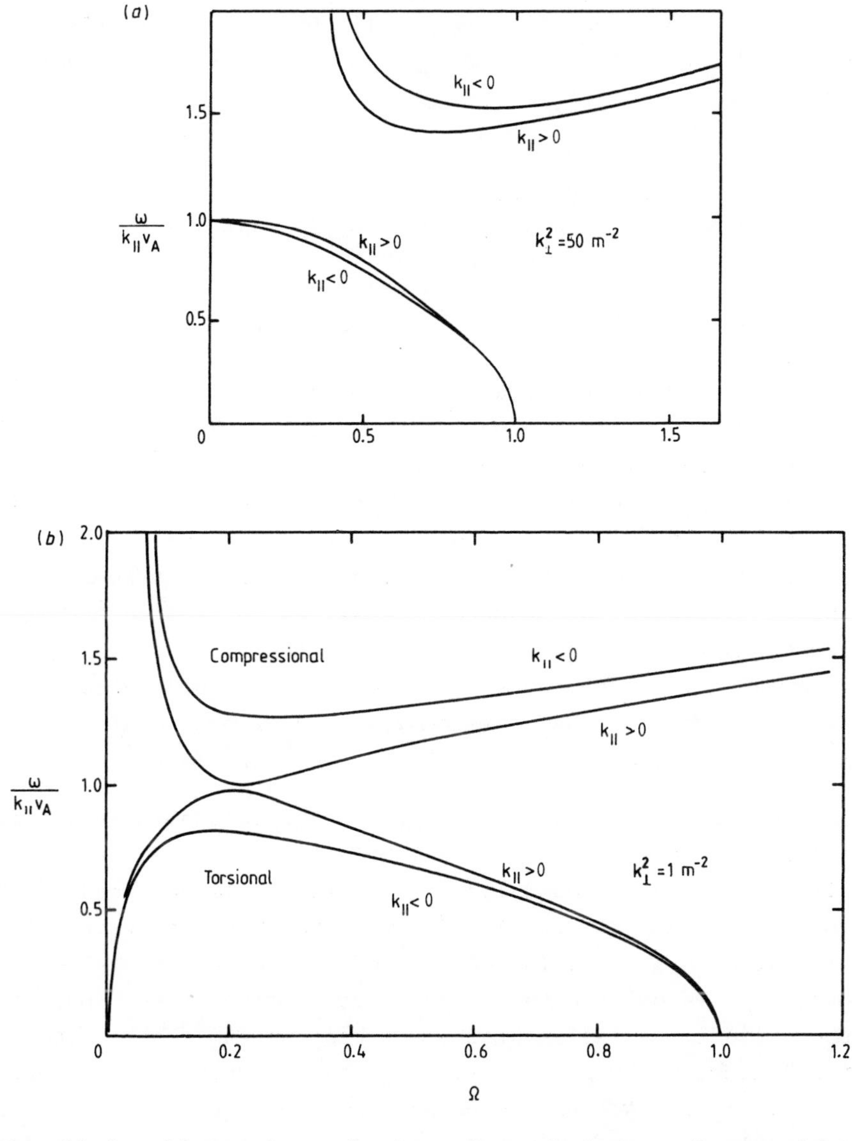

Fig. 11.2 Dispersion relations for a hydrogen plasma with $n_e = 2 \times 10^{19}\ m^{-3}$, D = 1.0$m^{-1}$, $k_{\perp}^2$ = 50m^{-2} or $k_{\perp}^2$ = 1m^{-2}.

conditions are considered in Sect. 11.4. It will be seen that the torsional and compressional wave types are both split into two modes having different phase velocities, $\omega/k_{\parallel}$, parallel to **B** depending on whether $k_{\parallel} > 0$ or $k_{\parallel} < 0$. This effect is especially pronounced at low $k_{\perp}$. The phase velocity, ω/k_z, in a direction parallel to B_z depends on m and q and can be determined with the aid of Eq. (11.5). This opens up a whole range of possibilities when it is realised that n, m and q are all small numbers of comparable magnitude and that n, m and q can all be positive or negative.

At high $k_{\perp}$, the dispersion relation is given closely by $F = 0$ since $k_{\perp} \to \infty$ as $F \to 0$. This relation describes high $k_{\perp}$ torsional waves whose phase velocity is given by

$$\omega/k_{\parallel} = v_A[1 - \Omega^2]^{\frac{1}{2}} \qquad (11.20)$$

In this case the group velocity vector is directed exactly parallel to **B** since $\partial\omega/\partial k_{\perp} = 0$. Consequently, an Alfven ray composed of a continuous spectrum of high $k_{\perp}$ components is guided along helical field lines, as shown in Fig. 6.6. This result has some interesting implications, discussed in detail in Ref. G1. For example, Alfven resonance surfaces in a toroidal plasma coincide with magnetic flux surfaces, even though surfaces of constant $k_{\parallel}v_A$ do not, since the mode converted wave energy is guided along the flux surface. It is apparent, therefore, that mode conversion of the compressional wave to the torsional wave does not require exact matching of $k_{\parallel}$ at all points in the resonance surface.

Rational flux surfaces are of special interest in toroidal plasmas. In a rational flux surface the steady field lines form closed loops of finite length and will therefore support localised standing torsional waves. The situation in a torus is complicated by the fact that B, and hence v_A, is not constant along a given field line. Due to the periodic variation of the Alfven speed along such a field line, there are two closely spaced resonance frequencies when an Alfven resonance surface coincides with a rational flux surface or when the length of the closed field lines is equal to an integer number of parallel wavelengths.

Wave propagation in a medium whose properties vary periodically with space or time is described by the Mathieu equation (G1). The above effects can be modelled experimentally by constructing a lumped transmission line, with periodically varying capacitors, and joined end to end to simulate a torus (G4).

At low wave frequencies the torsional wave has some peculiar and unexpected properties. Except in the limit $k_\perp \to 0$, the dispersion relation is $\omega = k_\parallel v_A$. This is as expected. The phase velocity in the z direction, from (11.5) is

$$v_z = \frac{\omega}{k_z} = \frac{B_z}{\left[\dfrac{B_o}{v_A} - \dfrac{mB_\theta}{\omega r}\right]} \qquad (11.21)$$

If all quantities here are positive, there is a $k_z = 0$ cutoff at

$$\omega = \omega_c = mv_A B_\theta/(rB_o) \qquad (11.22)$$

If $\omega > \omega_c$, v_z is positive and approaches v_A at high frequencies. If $\omega < \omega_c$, v_z is negative. If m or B_θ is negative there is no cutoff. These results appear to be totally uncharacteristic of the torsional wave. However they can be understood in terms of two simple properties of the torsional wave, (a) the phase velocity in a direction parallel to **B** is v_A and (b) as the wave propagates along **B**, the plane of polarisation rotates in space and time to follow the helical twist of the magnetic field lines. For example, consider a perturbation of the form

$$b = f(r)\cos(k_z z + m\theta - \omega t).$$

If $m > 0$, the global field pattern rotates clockwise in time at any given z but anticlockwise with increasing z at any given time. A maximum in b occurs at $\theta = (\omega t - k_z z)/m$. The plane of polarisation therefore rotates by this angle in a distance z or after a time t. During this time, a wavefront (or ray) will advance a distance $v_A t$ along a helical field line or a distance $z = v_A t B_z/B_o$ in the z direction. From Eq. (11.21) we find that $\theta = zB_\theta/(rB_z)$, which coincides with the twist in angle of a helical field line in a distance z. Since this angle is fixed by the equilibrium fields and is independent of wave frequency, the wave fronts may propagate along z with either $k_z > 0$ or $k_z < 0$ depending on the wave frequency.

11.2 Unstable modes

It was assumed above that **J** and **B** are parallel so that the plasma is in a force free equilibrium with $\mathbf{J} \times \mathbf{B} = 0$. If this equilibrium is disturbed by a small perturbation, the response of the plasma will be determined by the dispersion relation Eq. (11.16). This relation includes both torsional and compressional wave modes. It also includes solutions with real $k_\parallel$, real $k_\perp$ and with $\omega^2 < 0$. These are unstable modes with perturbations of the form

$$b = b(r)e^{\gamma t} \cos(k_z z + m\theta)$$

where $\omega^2 = -\gamma^2$. When $\gamma > 0$, the perturbation grows exponentially with time to a point where the linear theory is no longer valid. The subsequent behaviour becomes a nonlinear problem. It is possible, for example, that the perturbation may stop growing or "saturate", leading to a new, stable equilibrium.

The sign of m here determines the direction of rotation not with time but with distance. When B_θ and B_z are both positive, the equilibrium field lines twist clockwise as they advance in the positive z direction. A perturbation with $m/k_z > 0$ twists counter – clockwise, while $m/k_z < 0$ perturbations twist clockwise. It is the $m/k_z < 0$ perturbations which may be unstable, especially if the rate of twist or the helical pitch of the perturbation coincides with that of the equilibrium field. Such a perturbation is said to be resonant or *mode rational*.

The unstable solutions all have $|\omega| \ll \omega_{ci}$ and can therefore be described by the MHD ($\Omega = 0$) form of (11.16),

$$k_\perp^2 = \frac{(\omega^2/v_A^2 - k_\parallel^2)^2 - D^2 k_\parallel^2}{(\omega^2/v_A^2 - k_\parallel^2)}$$

with solutions

$$2\omega^2 = v_A^2(k_\perp^2 + 2k_\parallel^2) \pm v_A^2[k_\perp^4 + 4k_\parallel^2 D^2]^{\frac{1}{2}} \qquad (11.23)$$

If D = 0 (zero plasma current), the two solutions correspond to torsional and compressional waves. It is interesting to note that this dispersion relation only rarely appears in the literature on plasma instabilities. Stability problems are usually simplified by assuming (a) the plasma is incompressible, so the compressional wave cannot propagate or (b) the perturbations are force free, with **j** parallel to **B**. The latter assumption is not valid for either of the Alfven wave types.

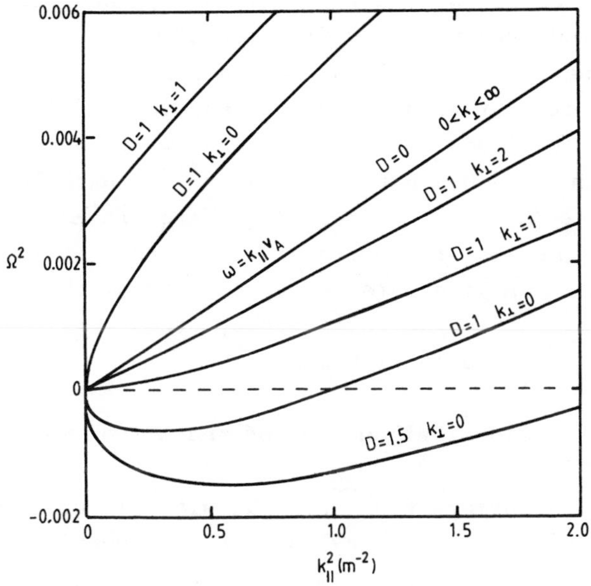

Fig. 11.3 Solutions of Eq.(11.23) showing the unstable kink modes ($\Omega^2 < 0$) and stable Alfven waves ($\Omega^2 > 0$) for a hydrogen plasma with $n_e = 2 \times 10^{19}$ m^{-3}. D = 0 corresponds to zero plasma current. Solutions above the line labelled $\omega = k_\parallel v_A$ are compressional waves. Solutions below the line are torsional waves or kink modes. The torsional wave evolves into a kink mode as D is increased.

Solutions of (11.23) are shown in Fig. 11.3 for several values of D and $k_\perp$. These values are typical of tokamak conditions. The vertical axis in Fig. 11.3 is the normalised frequency

$$\Omega^2 = \frac{\omega^2}{\omega_{ci}^{\ 2}} = \left[\frac{m_i}{\mu_o ne^2} \right] \frac{\omega^2}{v_A^{\ 2}}$$

The + sign in (11.23) corresponds to compressional wave solutions, while the − sign gives torsional wave or unstable modes. The solutions given in Fig. 11.3 may not satisfy boundary conditions. When boundary conditions are imposed, only certain values of $k_\perp$ are permitted, as described in Sect. 11.4. Nevertheless, it is clear that the most unstable solutions arise at low $k_\perp$ and when $D > 1\ m^{-1}$. Since the wave fields in the plasma vary as $J_m(k_\perp r)$, these fields are essentially independent of radius at low $k_\perp$. The whole plasma is therefore displaced as a finite $k_\parallel$ kink. This behaviour is closely related to the behaviour of low $k_\perp$, low ω torsional wave modes. From Eq. (11.23) we find that $k_\parallel = D$ when $k_\perp = 0$ and $\omega = 0$, corresponding to a parallel wavelength $\lambda_\parallel = 2\pi/k_\parallel = \ell/2 = 2\pi r B_z/B_\theta$ where ℓ is the pitch length of the helical field lines. This result imposes a maximum limit on the parallel wavelength of the torsional wave at low $k_\perp$. At longer wavelengths, the torsional wave evolves into an unstable kink mode.

11.3 Global eigenmodes of the Alfven wave

Consider a *homogeneous* plasma with a uniform field $\mathbf{B} = \hat{\mathbf{z}} B$ surrounded by a cylindrical conducting wall at $r = d$. In the *absence* of plasma current, such a system will support waveguide modes or eigenmodes of both the compressional and torsional Alfven waves. These modes are set up as a result of reflections at the wall, leading to standing waves in the radial and azimuthal directions. For example, the b_r wave field component for the first radial m = 0 mode is given by

$$b_r = b_o J_1(k_\perp r) \tag{11.24}$$

where $k_\perp d = 3.83$ to satisfy the boundary condition $b_r(d) = 0$. If $d = 0.2$m, $k_\perp = 19.15\ m^{-1}$. For any given ω, the dispersion relation, equation (10.9), will then yield a value of $k_\parallel$ for the compressional wave and a different $k_\parallel$ for the torsional wave.

Now suppose that the plasma density varies with radius. This case is considered analytically in Chapter 14 and has been analysed in detail by many authors (e.g. APPERT et al., E1). In this section we present a qualitative discussion of the results by assuming that each point in the plasma locally satisfies the homogeneous plasma dispersion relation. This is known as the WKB approximation. It is strictly only valid when the density varies by only a small amount in a distance equal to one wavelength in the radial direction. Nevertheless, the WKB approximation is very useful for qualitative discussion. This approximation works best for the high $k_\perp$ (small $\lambda_\perp$) waves described below.

In an *inhomogeneous*, *current-free* plasma, the compressional wave propagates as a waveguide mode with $k_\parallel$ ($= k_z$) being independent of radius. The phase velocity, $\omega/k_\parallel$, depends on the average Alfven speed and on the boundary conditions. For given ω and $k_\parallel$, the local value of $k_\perp^2$ varies with radius and is given by Eq. (10.7) for a current-free plasma. Typical $k_\perp^2(r)$ curves are shown in Fig. 11.4 for $\omega > \omega_{ci}$ or $\omega < \omega_{ci}$. When $\omega > \omega_{ci}$, the first radial $m = 0$ waveguide mode propagates with $\bar{k}_\perp d \sim 3.83$ where $\bar{k}_\perp$ is a radial average value of $k_\perp$.

When $\omega < \omega_{ci}$, it is seen from Fig. 11.4 that $k_\perp^2 = \infty$ at the radius where $F = 0$. This is known as the Alfven resonance layer. At this location, a torsional wave with $k_\perp^2 = \infty$ has the same ω and $k_\parallel$ as the compressional wave. The compressional wave propagates with $k_\perp^2 > 0$ on the high density side of the Alfven resonance layer. The torsional wave propagates on the low density side in a narrow layer where $k_\perp^2$ varies from 0 to ∞. In the MHD limit, this layer is infinitely thin, defining the surface where $\omega = k_\parallel v_A$. At finite frequency the layer is of finite width but the width usually remains only a small fraction of the plasma diameter. In the Alfven wave heating scheme, an evanescent

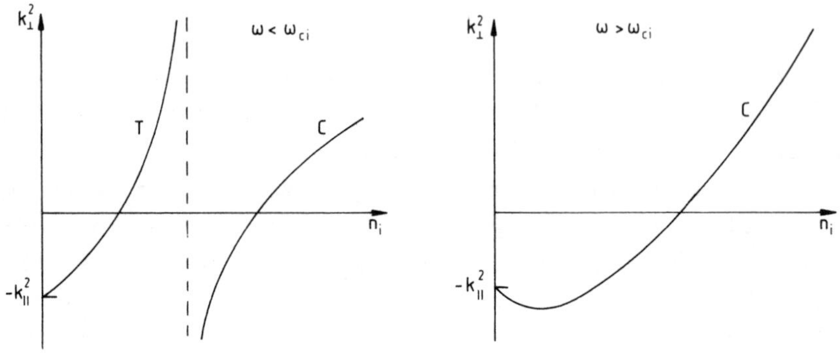

Fig. 11.4 $k_\perp{}^2$ vs ion density, n_i, in a current-free plasma for given ω and $k_\parallel$, when $\omega < \omega_{ci}$ or $\omega > \omega_{ci}$. T denotes torsional, C denotes compressional.

($k_\perp{}^2 < 0$) wave is generated at the plasma edge by an external antenna and mode converts to the torsional wave at the Alfven resonance layer. The torsional wave is then guided by the steady **B** field and heats the plasma by Landau damping on electrons near the resonance layer. The evanescent wave at the plasma edge is generally regarded as a continuation of the compressional wave branch, rather than the torsional wave branch, since it develops a large b_z component in the limit $n_i \to 0$. At higher n_i, the evanescent wave evolves into a torsional wave.

To examine the effects of plasma current in an inhomogeneous plasma, we can plot $k_\perp{}^2$ vs r using Eq.(11.16). In this case, a cylindrical eigenmode is defined by the integers n and m in Eq.(11.7) and $k_\parallel$ will then vary with r in general, depending on the q(r) profile.

Consider, for example, a current profile of the form

$$(\nabla \times \mathbf{B})_z = \frac{1}{r}\frac{\partial}{\partial r}(rB_\theta) = \mu_o J_z = \mu_o J_o(1 - r^2/a^2)^2 \qquad (11.25)$$

so

$$\frac{B_\theta}{r} = \frac{\mu_o J_o}{2}\left[1 - \frac{r^2}{a^2} + \frac{r^4}{3a^4}\right]$$

and

$$q(r) = \frac{rB_z}{RB_\theta} = \frac{q(a)}{3 - 3x^2 + x^4} \qquad (11.26)$$

where $x = r/a$, $q(a) = \dfrac{a\ B_z}{R\ B_\theta(a)}$, $B_\theta(a) = \dfrac{\mu_o J_o a}{6}$.

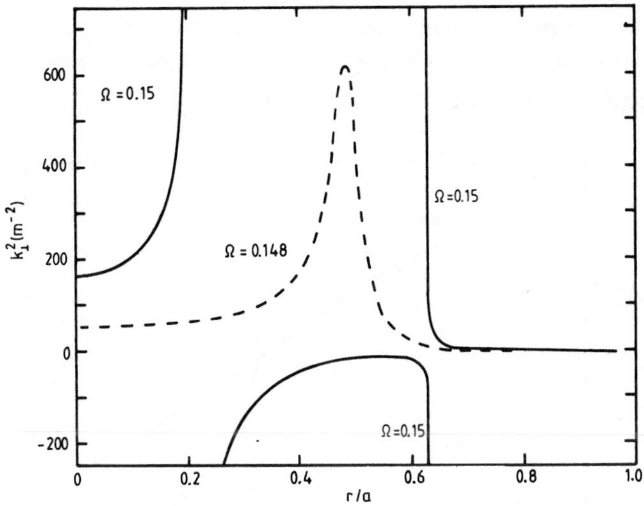

Fig. 11.5 $k_\perp^2$ vs r/a in hydrogen, m = +2, n = +1, q(a) = 3.

Fig. 11.5 shows $k_\perp^2$ vs r/a in a hydrogen plasma with $n_i = n_o(1 - x^2)$, $n_o = 2 \times 10^{19}\ m^{-3}$, R = 1.0 m, and q(a) = 3. A wide variety of $k_\perp^2(r)$ diagrams can be constructed, most of which are very sensitive to the assumed n(r) and q(r) profiles and to the signs of m and n. The most important feature is the location of the Alfven resonance layer, where $k_\perp^2 = \infty$, given from equations (11.7) and (11.16), by

$$\left(n + \frac{m}{q}\right)^2 = Sn_i \qquad S = \frac{\Omega^2 \mu_o e^2 R^2}{(1 - \Omega^2)m_i} \qquad (11.27)$$

If $y(r) = (n + m/q)^2$ and $z(r) = Sn_i$ then the resonance layer is located at the radius (or radii) where the y(r) and z(r) profiles intercept. Typical y(r) and z(r) profiles are given in Fig. 11.6 for the same parameters as those in Fig. 11.5. The most interesting profiles are those where the y(r) and z(r) profiles are very similar functions of r, and where $k_\perp^2$ is large over a significant fraction of the plasma cross section. This is the case shown in Figs. 11.5 and 11.6 at $\Omega = 0.148$ for the n = +1, m = +2 mode. Provided that the y and z profiles don't intersect, there is no resonance layer in the plasma. Provided also that $\bar{k}_\perp > 20\ m^{-1}$, then an eigenmode (or global or discrete mode) of the torsional wave will satisfy the boundary conditions. These conditions can occur at a frequency just below the minimum frequency for an Alfven resonance to appear in the plasma, the so-called "lower edge of the Alfven continuum".

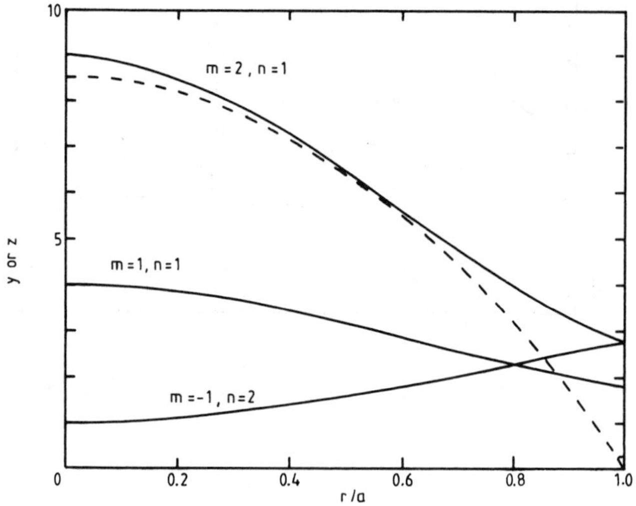

Fig. 11.6 Profiles of y (solid curves) and z (dashed curve, $\Omega = 0.148$) with q(a) = 3 and $n_i = n_o(1-r^2/a^2)$.

The average value of $k_\perp$ can exceed $20\ m^{-1}$ for any plasma conditions one choses, provided the wave frequency is chosen so that $k_\perp \to \infty$ at some point in the plasma. However, for realistic wave damping, $k_\perp$ will never be infinite and one cannot expect to encounter a global mode for arbitrary $n_i(r)$ and q(r) profiles. The most widely studied global modes (F4) are those occuring in plasmas where $n_i(r)$ is close to parabolic and where $q(a) \sim 3 \times q(0)$. Global modes are encountered for those conditions where $k_{\parallel}v_A$ is approximately independent of radius.

The $k_{\perp}^2$ profile at $\Omega = 0.148$ in Fig. 11.5 is a little too narrow to support an eigenmode. However, a very slight change in the n_e or J profile could enhance $k_{\perp}^2$ over a wider region of the plasma, allowing an eigenmode to form. Note that if the frequency is raised slightly to $\Omega = 0.15$ for the conditions shown in Fig. 11.5, two Alfven resonance layers appear in the plasma. It is easy to see that the y and z profiles intercept at two values of r when $\Omega > 0.148$ since this is equivalent to shifting the dashed z profile in Fig. 11.6 slightly upwards.

Global torsional wave modes were predicted (APPERT et. al, E1) and observed (F6) almost simultaneously. They are only weakly damped since there is no Alfven resonance layer. Since they exist in such a narrow frequency range which varies with the density, they are of no real interest for wave heating. However, the global modes are useful as a diagnostic of plasma conditions (COLLINS et al., F4).

Given that wave energy in the torsional wave propagates almost entirely along B lines, it is not immediately obvious how global modes can be excited under the usual experimental conditions where the antenna is located at the plasma edge. There is no question that they are, but the mechanism is not clear. For example, in the experiments described by COLLINS et al.(F4), global modes were excited by unshielded, external antennas. Electrostatic coupling or radial current flow through the plasma may then be important. Alternatively, non-resonant mode conversion of compressional wave energy may be involved. An off-resonance response can be observed in any resonant system when realistic damping processes are included, since the system then has a finite quality factor or a finite bandwidth. The global torsional wave modes appear under conditions where the resonance condition is almost satisfied over most of the plasma cross-section.

11.4 Surface waves in a current–carrying plasma

As shown in Fig. 11.2, a steady plasma current has only a weak effect on high $k_\perp$ Alfven wave modes but a strong effect on the dispersion properties of low $k_\perp$ wave modes. As a result, Alfven surface waves are strongly affected by a steady plasma current, especially at low wave frequencies. In this Section, the dispersion relation for surface waves in a cylindrical current–carrying plasma is derived for a relatively simple case where (a) the plasma density and the plasma current density are independent of r between r = 0 and r = a, and (b) the plasma is surrounded by a vacuum in the region $a < r < \infty$.

The wave electric and magnetic fields in the plasma are given by Eqs. 11.9, 11.11, 11.14, 11.15 and 11.19. The wave fields in the vacuum cannot simply be obtained, as in Ch. 8, by setting $v_A = \infty$ in the expressions for the wave fields in the plasma. In the vacuum region, the zero order B_θ field varies inversely as r, and hence $k_\parallel$ is a strong function of r. The above expressions for the wave fields in the plasma were derived on the assumption that $k_\parallel$ was independent of r. In order to compute the vacuum fields, it is more convenient to use the conventional r, θ, z coordinate system, and assume perturbations of the form $f(r)\exp[i(k_z z + m\theta - \omega t)]$. Since $E_\parallel$ is continuous at the plasma–vacuum boundary, and since $E_\parallel = 0$ in the plasma (Eq. 11.9), it is necessary to allow for finite E_z and E_θ components in the vacuum. The vacuum and plasma wave fields can be matched at the boundary using the relations

$$\left.\begin{aligned}
\mathbf{E}\cdot\mathbf{B} &= E_\parallel B_o = E_\theta B_\theta + E_z B_z = 0 \\
(\mathbf{E}\times\mathbf{B})_r &= E_\perp B_o = E_\theta B_z - E_z B_\theta \\
\mathbf{b}\cdot\mathbf{B} &= b_\parallel B_o = b_\theta B_\theta + b_z B_z \\
(\mathbf{b}\times\mathbf{B})_r &= b_\perp B_o = b_\theta B_z - b_z B_\theta
\end{aligned}\right\} \qquad (11.28)$$

The vacuum wave fields are described by Maxwell's equations $\nabla\times\mathbf{E} = i\omega\mathbf{b}$ and $\nabla\times\mathbf{b} = 0$. The components of these equations yield

$$\left.\begin{aligned} b_r &= \left(\frac{1}{ik_z}\right)\frac{\partial b_z}{\partial r} & \qquad E_r &= \left(\frac{\omega m}{k_z{}^2 r}\right) b_z + \left(\frac{1}{ik_z}\right)\frac{\partial E_z}{\partial r} \\ b_\theta &= \left(\frac{m}{k_z r}\right) b_z & \qquad E_\theta &= \left(\frac{m}{k_z r}\right) E_z + \left(\frac{i\omega}{k_z{}^2}\right)\frac{\partial b_z}{\partial r} \end{aligned}\right\} \quad (11.29)$$

The E_z and b_z components are collected on the right hand sides. The general solution for the vacuum fields can be expressed as a sum of the transverse electric ($E_z = 0$) and transverse magnetic ($b_z = 0$) modes in the vacuum. Since

$$\nabla\cdot\mathbf{b} = \frac{1}{r}\frac{\partial(rb_r)}{\partial r} + \frac{im}{r} b_\theta + ik_z b_z = 0$$

where b_r and b_θ are given by Eq.(11.29), the TE mode is described by

$$\frac{\partial^2 b_z}{\partial r^2} + \frac{1}{r}\frac{\partial b_z}{\partial r} - \left[k_z{}^2 + \frac{m^2}{r^2}\right] b_z = 0 \qquad (11.30)$$

$$\therefore \qquad b_z = b_v K_m(k_z r) \qquad (11.31)$$

The I_m solution is ignored since $I_m \to \infty$ as $r \to \infty$. Similarly, $\nabla\cdot\mathbf{E} = 0$ in the vacuum, and the TM mode is described by the equation

$$\frac{\partial^2 E_z}{\partial r^2} + \frac{1}{r}\frac{\partial E_z}{\partial r} - \left[k_z{}^2 + \frac{m^2}{r^2}\right] E_z = 0 \qquad (11.32)$$

$$\therefore \qquad E_z = E_v\, K_m(k_z r) \qquad (11.33)$$

We assume that b_r and $b_\parallel$ are continuous at the plasma–vacuum boundary, $r = a$. Any $j = \infty$ surface current must be parallel to $\mathbf{B}$, otherwise an infinite $\mathbf{j} \times \mathbf{B}$ force would be generated at the boundary. This means that $b_\parallel$ is continuous, but not $b_\perp$. From 11.8, 11.14, 11.29 and 11.31, continuity of b_r, at $r = a$, requires that

$$\frac{ik_\parallel}{k_\perp^2}\left(\frac{mG}{aF}b_\parallel + \frac{\partial b_\parallel}{\partial r}\right) = \frac{b_v}{ik_z} K_m'(k_z a) \tag{11.34}$$

Continuity of $b_\parallel$ yields

$$b_o J_m(k_\perp a) = \frac{1}{B_o}\left(B_z + \frac{mB_\theta(a)}{k_z a}\right) b_v K_m(k_z a)$$

or

$$\frac{b_o}{b_v} = \frac{k_\parallel K_m(k_z a)}{k_z J_m(k_\perp a)} \tag{11.35}$$

The derivatives of J_m and K_m are given by

$$J_m'(k_\perp r) = k_\perp J_{m-1}(k_\perp r) - \frac{m}{r} J_m(k_\perp r)$$

$$K_m'(k_z r) = - k_z K_{m-1}(k_z r) - \frac{m}{r} K_m(k_z r)$$

where $J_{-m}(x) = (-1)^m J_m(x) \qquad K_{-m}(x) = K_m(x)$

Equations (11.34) and (11.35) yield

$$\frac{k_\parallel^2}{k_\perp^2}\left[m\left(\frac{G}{F} - 1\right) + \frac{xJ_{m-1}(x)}{J_m(x)}\right] = \left[\frac{yK_{m-1}(y)}{K_m(y)} + m\right] \tag{11.36}$$

where $x = k_\perp a$ and $y = k_z a$

Solutions with $k_\perp^2 < 0$ can also be found, in which case the solution of Eq.(11.17) is $b_\parallel = b_o I_m(k_s r)$ where $k_s^2 = -k_\perp^2$. Continuity of b_r and $b_\parallel$ in this case yields

$$\frac{k_\parallel^2}{k_s^2}\left[m\left(\frac{G}{F} - 1\right) + \frac{zI_{m-1}(z)}{I_m(z)}\right] = -\left[\frac{yK_{m-1}(y)}{K_m(y)} + m\right] \tag{11.37}$$

where $z = k_s a$ and $y = k_z a$

and $I_m'(k_s r) = k_s I_{m-1}(k_s r) - \frac{m}{r} I_m(k_s r)$

(a) MHD solutions

The essential features of surface wave modes in a current carrying plasma can be extracted from (11.34) and (11.35) by examining the solutions at low wave frequencies where $\omega \ll \omega_{ci}$. In this limit, the parallel and perpendicular wavelengths for the surface wave modes are much larger than the plasma radius, a, so $x \ll 1$, $y \ll 1$ and $z \ll 1$. The Bessel functions appearing in (11.36) and (11.37) can then be expressed in the form

$$\frac{x\, J_{m-1}(x)}{J_m(x)} = m + |m| \qquad \text{when } x \ll 1$$

$$\frac{y\, K_{m-1}(y)}{K_m(y)} = |m| - m \qquad \text{when } y \ll 1$$

$$\frac{z\, I_{m-1}(z)}{I_m(z)} = |m| + m \qquad \text{when } z \ll 1$$

Using these approximations, equation (11.36) reduces to

$$k_\parallel^2 \left(\frac{mG}{F} + |m| \right) = k_\perp^2 |m|$$

where $k_\perp^2 = (F^2 - G^2)/F$, $F = (\omega^2/v_A^2) - k_\parallel^2$, $G = -Dk_\parallel$.

Hence
$$\left.\begin{aligned} 2k_\parallel^2 - Dk_\parallel - (\omega^2/v_A^2) &= 0 \qquad \text{if} \quad m > 0 \\ 2k_\parallel^2 + Dk_\parallel - (\omega^2/v_A^2) &= 0 \qquad \text{if} \quad m < 0 \end{aligned}\right\} \quad (11.38)$$

Equation (11.38) is the dispersion relation for surface wave modes. Exactly the same relation is obtained, when $k_\perp^2 < 0$, from Eq.(11.37). Solutions of (11.38) are given in Fig. 11.7 for a hydrogen plasma with $n_e = 2.08 \times 10^{19}$ m^{-3} ($\omega_{ci}/v_A = 20$) and $D = 1.0 m^{-1}$, typical of tokamak conditions. Solutions are plotted up to $\omega/\omega_{ci} = 0.2$. When $\omega > 0.2\omega_{ci}$, Eq. (11.38) is not valid since the MHD approximation $\omega \ll \omega_{ci}$ is invalid and since the approximations $x \ll 1$, $y \ll 1$ are invalid.

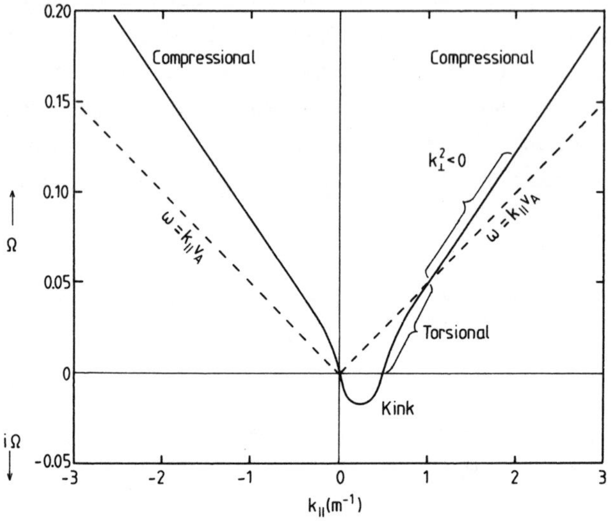

Fig. 11.7 Dispersion relation for $m > 0$ surface waves in a current–carrying hydrogen plasma, at low wave frequencies, with $D = 1.0\ m^{-1}$, $n_e = 2.08 \times 10^{19}\ m^{-3}$.

We note the following features of Fig. 11.7 and Eq.(11.38):

□ Each of the dispersion relations in (11.38) is quadratic in $k_\parallel$ for any given ω, so that there are two surface wave modes, one with $k_\parallel > 0$ and one with $k_\parallel < 0$. Modes with $m > 0$ and $m < 0$ are mirror images with respect to the signs of $k_\parallel$. A mode with $m > 0$ and $k_\parallel > 0$ has the same dispersion relation as a mode with $m < 0$ and $k_\parallel < 0$. A mode with $mk_\parallel < 0$ is well behaved, but a mode with $mk_\parallel > 0$ has some pecular properties associated with the fact that its dispersion relation intersects the dispersion relation for torsional waves at $k_\parallel = D$ (when $m > 0$) or $k_\parallel = -D$ (when $m < 0$).

□ At $\omega = 0$, the solutions of 11.38 are $k_\parallel = 0$ or $k_\parallel = D/2$ if $m > 0$ and $k_\parallel = -D/2$ if $m < 0$. When $\omega = 0$ and $k_\parallel = D/2$, $k_\perp{}^2 = D^2 - k_\parallel{}^2$ so $k_\perp{}^2 = 3D^2/4$. It is unusual for a wave to have a finite wavenumber when $\omega \to 0$. As $\omega \to 0$ the parallel wavelength tends to a limiting value $\lambda_\parallel = 2\pi/k_\parallel = \ell$ where ℓ is the pitch length of the helical field lines. In this respect, the surface wave mode behaves in a similar way to the torsional wave, as described in Section 11.2. In fact, for $k_\parallel$ in the range $D/2 < k_\parallel < D$, the surface wave IS a torsional wave. One can identify the wave as torsional from Eq.(11.23) and from the discussion at the end of Section 11.2 regarding the signs in equation (11.23).

- The two surface wave solutions shown in Fig. 11.7 are connected, at $\omega = 0$, by the unstable kink mode. The kink mode exists with finite $k_\parallel$ but with $\Omega^2 < 0$.

- At high wave frequencies, where $k_\parallel \gg D/2$, Eq.(11.38) reduces to the form $\omega/k_\parallel = \sqrt{2}\, v_A$ for all $m \neq 0$ surface waves. This result agrees with the dispersion relation, Eq.(9.22), derived previously for zero plasma current.

- The surface wave dispersion relation intersects the torsional wave dispersion relation $\omega = \pm k_\parallel v_A$ when $k_\parallel = D$ $(m > 0)$ or $k_\parallel = -D$ $(m < 0)$. Since $F = 0$ under these conditions, $k_\perp{}^2 = \pm \infty$ at the point of intersection. However, the approximation $x \ll 1$ breaks down near the point of intersection, and valid solutions of (11.36) and (11.37) must be obtained by using more exact expressions for the Bessel functions. These solutions are described by CRAMER and DONNELLY (D3).

- As shown in Fig. 11.7, $k_\perp{}^2 < 0$ in the range $D < k_\parallel < 2D$. At $k_\parallel = 2D$, $k_\parallel{}^2 = 0$ while at $k_\parallel = D$, $k_\perp{}^2 = -\infty$.

(b) Finite frequency solutions

Dispersion relations for surface waves at finite frequency can be obtained by numerical integration of Eqs.(11.17) and (11.30), using continuity of b_r and $b_\parallel$ to restart the integration at the plasma-vacuum boundary. It is also convenient, for the numerical calculations, to terminate the integration at a conducting wall boundary, rather than integrate to $r = \infty$. Solutions of these equations are given in Fig. 11.8. The b_z and b_θ components in the plasma are given, from equation (11.28), by

$$b_z = (b_\parallel B_z - b_\perp B_\theta)/B_0 \qquad \text{and} \qquad b_\theta = (b_\parallel B_\theta + b_\perp B_z)/B_0$$

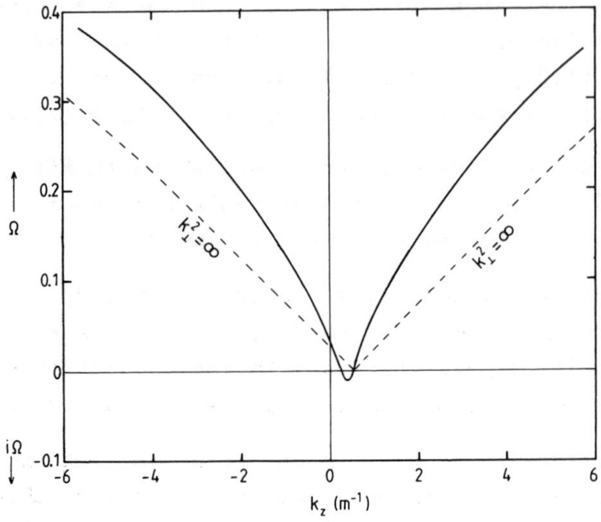

Fig. 11.8 Dispersion relation for an m = −1 surface wave in a hydrogen plasma with $n_e = 2.08 \times 10^{19}$ m^{-3}, a = 0.1 m d = 0.15 m, R = 0.5 m, q = 4.

It will be seen from Fig. 11.8 that the asymmetry between $k_{\parallel} > 0$ and $k_{\parallel} < 0$ modes is less pronounced at high wave frequencies than at low frequencies. The solutions at high frequencies are similar to those described in Section 10.5, indicating that the plasma current does not play a very significant role at high frequencies. This is consistent with the results shown in Fig. 11.2 where the differences between $k_{\parallel} > 0$ and $k_{\parallel} < 0$ modes remain small at high $k_{\perp}$ and at high frequencies. For surface wave modes, $k_{\perp}$ increases with wave frequency, and the mode evolves into a conventional "body" wave.

12. THE DIELECTRIC TENSOR

Wave propagation in a plasma with a single ion species has been described in the previous Chapters in terms of the generalised Ohm's law for the plasma. An alternative description of plasma waves can be given in terms of the dielectric tensor. Even though the word "tensor" will send shivers down the spines of many graduate students, the dielectric tensor for a cold, current-free plasma is easily derived and is extremely useful. The dispersion relation for waves in a plasma with one or more ion species is most widely quoted in terms of the dielectric tensor.

For a dielectric medium it is usual to define the electric displacement, in SI units, by $\mathbf{D} = \epsilon_0\mathbf{E} + \mathbf{P} = K\epsilon_0\mathbf{E}$ where K is the dielectric constant and $\mathbf{P}$ is the polarisation. For a plasma, it is convenient to include $\mathbf{j}$ as part of the total displacement current and to define the dielectric tensor $\underline{\mathbf{K}}$ as follows, with $\partial/\partial t = -i\omega$:

$$\nabla \times \mathbf{b} = \mu_0(\mathbf{j} - i\omega\epsilon_0\mathbf{E}) = -i\omega\mu_0\mathbf{D} \tag{12.1}$$

where
$$\mathbf{D} = \epsilon_0\mathbf{E} - \mathbf{j}/(i\omega) = \epsilon_0\,\underline{\mathbf{K}} \cdot \mathbf{E} \tag{12.2}$$

Note that $\underline{\mathbf{K}} \cdot \mathbf{E}$ is a vector. The product of $\underline{\mathbf{K}}$ and $\mathbf{E}$ is defined by Eq. (12.6) below. For those readers not familiar with tensors, a simple example will be useful. Consider a semiconductor in a magnetic field $\mathbf{B} = \hat{\mathbf{z}}B$, where Ohm's law has the form $\mathbf{E} = \eta\mathbf{j} + \mathbf{j} \times \mathbf{B}/(ne)$. Then

$$E_x = \eta j_x + (B/ne)j_y, \quad E_y = -(B/ne)j_x + \eta j_y \quad \text{and} \quad E_z = \eta j_z .$$

These relations can be expressed in matrix form as

$$\mathbf{E} = \underline{\mathbf{R}} \cdot \mathbf{j} = \begin{bmatrix} \eta & B/ne & 0 \\ -B/ne & \eta & 0 \\ 0 & 0 & \eta \end{bmatrix} \begin{bmatrix} j_x \\ j_y \\ j_z \end{bmatrix}$$

where $\underline{\mathbf{R}}$ is the resistivity tensor. It has 9 elements in this case. The diagonal elements are all equal to η. The off-diagonal elements, ±B/ne are known as the Hall resistivity. The dielectric tensor for a plasma also has nine elements, but the elements are more complicated.

12.1 Plasma Equations

Consider a plasma containing several ion species of number density n_k and charge $q_k = Z_k e$. For electrons, $Z_k = -1$. The equation of motion for each species, ignoring collisions, and the total current density are given by

$$m_k \frac{\partial \mathbf{v}_k}{\partial t} = q_k(\mathbf{E} + \mathbf{v}_k \times \mathbf{B}) \qquad (12.3)$$

$$\mathbf{j} = \sum_k n_k\, q_k\, \mathbf{v}_k \qquad (12.4)$$

Let $\mathbf{B} = \hat{\mathbf{z}}B$. For each species, dropping subscript k,

$$-i\omega m\, v_x = q(E_x + v_y B)$$

$$-i\omega m\, v_y = q(E_y - v_x B)$$

$$-i\omega m\, v_z = qE_z$$

Let $\omega_c = qB/m$ (ω_c is the cyclotron frequency, $q = -e$ for electrons). Solving for v_x and v_y, we obtain

$$\left.\begin{aligned} v_x &= \frac{\omega^2}{B(\omega^2 - \omega_c{}^2)}\left[\frac{i\omega_c}{\omega}E_x - \frac{\omega_c{}^2}{\omega^2}E_y\right] \\ v_y &= \frac{\omega^2}{B(\omega^2 - \omega_c{}^2)}\left[\frac{\omega_c{}^2}{\omega^2}E_x + \frac{i\omega_c}{\omega}E_y\right] \\ v_z &= i\omega_c E_z/(\omega B) \end{aligned}\right\} \qquad (12.5)$$

Let $\omega_p{}^2 = nq^2/(\epsilon_0 m)$ where n is the number density (not the refractive index), ω_p is the plasma frequency for each species, and let

$$P = 1 - \frac{1}{\omega^2}\sum_k \omega_p^{\,2} \qquad (P = \text{Plasma})$$

$$R = 1 - \sum_k \frac{\omega_p^{\,2}}{\omega(\omega + \omega_c)} \qquad (R = \text{Right})$$

$$L = 1 - \sum_k \frac{\omega_p^{\,2}}{\omega(\omega - \omega_c)} \qquad (L = \text{Left})$$

$$S = \frac{R + L}{2} = 1 - \sum_k \frac{\omega_p^{\,2}}{(\omega^2 - \omega_c^{\,2})} \qquad (S = \text{Sum})$$

$$D = \frac{R - L}{2} = \sum_k \frac{\omega_p^{\,2}\omega_c}{\omega(\omega^2 - \omega_c^{\,2})} \qquad (D = \text{Difference})$$

Eqs. (12.2), (12.4) and (12.5) are easily combined to give

$$\begin{aligned}(\underline{\mathbf{K}} \cdot \mathbf{E})_x &= SE_x - iDE_y \\ (\underline{\mathbf{K}} \cdot \mathbf{E})_y &= SE_y + iDE_x \\ (\underline{\mathbf{K}} \cdot \mathbf{E})_z &= PE_z\end{aligned}$$

which can be expressed in matrix form as

$$\underline{\mathbf{K}} \cdot \mathbf{E} = \begin{bmatrix} K_{xx} & K_{xy} & 0 \\ K_{yx} & K_{yy} & 0 \\ 0 & 0 & K_{zz} \end{bmatrix} \begin{bmatrix} E_x \\ E_y \\ E_z \end{bmatrix} \qquad (12.6)$$

where

$K_{xx} = K_{yy} = K_\perp = S, \quad K_{zz} = K_\parallel = P \quad$ and $\quad K_{yx} = -K_{xy} = iD.$

12.2 Dispersion Relations

Consider plane waves of the form

$$\mathbf{E} = E_o e^{i(\mathbf{k}\cdot\mathbf{r} - \omega t)} = E_o e^{i(k_x x + k_y y + k_z z - \omega t)}$$

and define the refractive index $\mathbf{n} = c\mathbf{k}/\omega$ having magnitude $n = c/v$ where $v = \omega/k$ is the phase velocity. Using Maxwell's $\nabla \times \mathbf{E} = i\omega\mathbf{b}$ equation and the notation $\nabla = i\mathbf{k}$,

$$\nabla \times (\nabla \times \mathbf{E}) = -\mathbf{k} \times (\mathbf{k} \times \mathbf{E}) = \frac{\omega^2}{c^2} \underline{\mathbf{K}} \cdot \mathbf{E}$$

so

$$\boxed{\mathbf{n} \times (\mathbf{n} \times \mathbf{E}) = -\underline{\mathbf{K}} \cdot \mathbf{E}} \qquad (12.7)$$

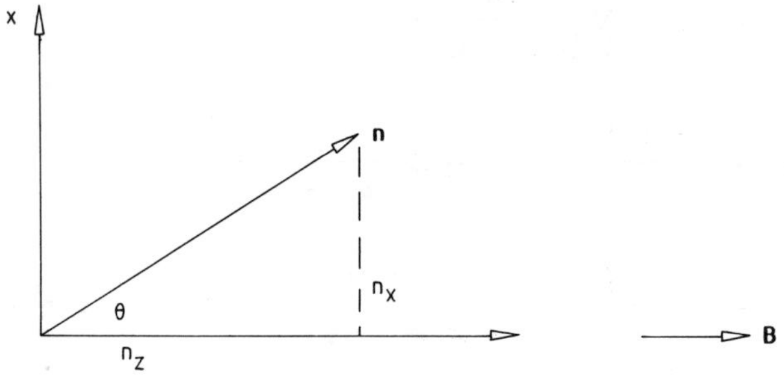

Fig. 12.1 Orientations of **n** and **B**

Let **n** and **k** lie in the x–z plane so $n_y = k_y = 0$, as shown in Fig. 12.1. The components of $\mathbf{n} \times (\mathbf{n} \times \mathbf{E})$ are

$$\mathbf{n} \times (\mathbf{n} \times \mathbf{E}) = \begin{bmatrix} -n_z^2 & 0 & n_x n_z \\ 0 & -(n_x^2+n_z^2) & 0 \\ n_x n_z & 0 & -n_x^2 \end{bmatrix} \begin{bmatrix} E_x \\ E_y \\ E_z \end{bmatrix}$$

From equations (12.6) and (12.7) we obtain

$$\left.\begin{aligned} (S - n_z^2)E_x - iDE_y + n_x n_z E_z &= 0 \\ iDE_x + (S - n_x^2 - n_z^2)E_y &= 0 \\ n_x n_z E_x + (P - n_x^2)E_z &= 0 \end{aligned}\right\} \qquad (12.8)$$

The solution of this set, given that $S^2 - D^2 = LR$, is

$$Sn_x^4 + [n_z^2(P+S)-LR-SP]n_x^2 + P[LR-2Sn_z^2+n_z^4] = 0 \qquad (12.9)$$

Alternatively, since $n_x = n \sin\theta$ and $n_z = n \cos\theta$ where θ is the angle between **n** and **B**,

$$\boxed{A_1 n^4 - B_1 n^2 + C_1 = 0} \tag{12.10}$$

where
$$A_1 = S \sin^2\theta + P \cos^2\theta$$
$$B_1 = LR \sin^2\theta + SP(1 + \cos^2\theta)$$
$$C_1 = LRP$$

Equation (12.10) can also be expressed as

$$\boxed{P k_{\parallel}^4 + Q k_{\parallel}^2 + M = 0} \tag{12.11}$$

where $$Q = (P + S)k_{\perp}^2 - \frac{2\omega^2 PS}{c^2}, \qquad k_{\parallel} = k_z ,$$

and $$M = S k_{\perp}^4 - \frac{\omega^2}{c^2}(LR + SP)k_{\perp}^2 + \frac{\omega^4 PLR}{c^4}$$

Eqs. (12.9), (12.10) and (12.11) are all equivalent forms of the cold plasma dispersion relation including finite electron mass, the Hall term in Ohm's law, the vacuum displacement current and one or more ion species. As such, the dispersion relation describes all known cold plasma wave types, including the torsional and compressional Alfven waves. For each wave type the ratio, G_V, of the group velocity components can be determined from equations (5.3) and (5.4):

$$G_V = \frac{\partial\omega/\partial k_{\perp}}{\partial\omega/\partial k_{\parallel}} = \frac{k_{\perp}[2Sk_{\perp}^2 + (P+S)k_{\parallel}^2 - \omega^2(LR+SP)/c^2]}{k_{\parallel}(Q + 2Pk_{\parallel}^2)} \tag{12.12}$$

The wave electric field components can be found from (12.8). The wave magnetic field components follow from Maxwell's $\nabla \times \mathbf{E}$ equation. The dispersion relation with finite k_y or n_y is obtained by replacing n_x^2 by $n_{\perp}^2$ in equation (12.9) where $n_{\perp}^2 = n_x^2 + n_y^2$. Alternatively, $k_{\perp}^2 = k_x^2 + k_y^2$ in (12.11).

12.3 Low frequency approximations

We now show that the dispersion relations (10.7) and (12.11) are equivalent at low wave frequencies where $\omega \ll \omega_{pe}$ and $\omega \ll \omega_{ce}$. For a single ion species plasma, $n_e = Z_i n_i$, $q_i = Z_i e$

$$\frac{\omega_{pi}^2}{\omega_{ci}} = -\frac{\omega_{pe}^2}{\omega_{ce}} = \frac{c^2\omega_{ci}}{v_A^2}$$

$$v_A^2 = \frac{B^2}{\mu_o n_i m_i} \qquad \omega_{ce} = -\frac{Be}{m_e} \qquad c^2 = \frac{1}{\mu_o \epsilon_o}$$

Defining $\Omega_i = \omega/\omega_{ci}$ and $\Omega_e = -\omega/\omega_{ce}$,

$$P = -\frac{\omega_{pe}^2}{\omega^2} = -\frac{c^2}{v_A^2\,\Omega_e\Omega_i}$$

$$R = -\frac{1}{\omega}\left[\frac{\omega_{pe}^2}{\omega_{ce}} + \frac{\omega_{pi}^2}{(\omega + \omega_{ci})}\right] = \frac{c^2}{v_A^2(1 + \Omega_i)}$$

$$L = \frac{1}{\omega}\left[\frac{\omega_{pe}^2}{\omega_{ce}} - \frac{\omega_{pi}^2}{(\omega - \omega_{ci})}\right] = \frac{c^2}{v_A^2(1 - \Omega_i)}$$

$$S = -\frac{\omega_{pi}^2}{(\omega^2 - \omega_{ci}^2)} = \frac{c^2}{v_A^2(1 - \Omega_i^2)}$$

$$D = -\frac{c^2\Omega_i}{v_A^2(1 - \Omega_i^2)}$$

The dispersion relation (12.11) can be further simplified when $\Omega_e \ll 1$ since then $S \ll P$ and $LR \ll SP$. If we also ignore very large values of $k_\perp$ so that $Sk_\perp^4/P \ll k_\parallel^4$, then it is easy to show that (12.11) reduces to Eq.(10.7).

Equation (10.7) was derived by ignoring displacement current and finite electron mass (in Ohm's law). The displ-acement current is responsible for the unity terms in P, R, L and S. At frequencies near or less than the ion cyclotron frequency, the displacement current is important only at extremely low plasma densities, typically $< 10^{13}$ m^{-3}, where

v_A approaches or exceeds c. Finite electron mass is important only at large $k_\perp$ (and therefore only for the torsional wave). Finite plasma temperature effects are more important than finite m_e at high $k_\perp$ if the plasma beta, β, exceeds m_e/m_i, and are considered in Section 12.7.

These effects are illustrated in Fig. 12.2 by comparing solutions of Eqs.(10.7) and (12.11) for fixed ω and $k_\parallel$ and plotting $k_\perp^2$ vs n_e. This is a common method of showing the variation of $k_\perp^2$ for a wave of given ω and $k_\parallel$ propagating in a plasma whose density varies with radius. The results in Fig. 12.2 also highlight the fact that at some radius in the plasma the torsional wave may also propagate with the same ω and $k_\parallel$ as the compressional wave, as described in Sec. 11.3. In Fig. 12.2, the solution at high $k_\perp$ and low n_e corresponds to a torsional wave with small perpendicular wavelength. The solution labelled $k_\perp^2 = -k_\parallel^2$ at low n_e persists at $\omega > \omega_{ci}$ (see Fig. 11.4) and is an evanescent branch of the compressional wave, having a large b_z component when $n_e \to 0$.

12.4 High $k_\perp$ approximation

The behaviour of the torsional wave at high $k_\perp$ can be deduced from Eq. (12.9) by ignoring the last term, assuming that n_x^2 and n_x^4 terms are more significant. Assuming also that $\omega \ll \omega_{pe}$ and $\omega \ll \omega_{ce}$, and replacing k_x^2 by $k_\perp^2$, we find that

$$k_\perp^2 = \frac{1}{\Omega_e \Omega_i}\left[k_\parallel^2(1 - \Omega_i^2) - \frac{\omega^2}{v_A^2}\right] \tag{12.13}$$

This relation is also plotted in Fig. 12.2 and is seen to be a good approximation at high $k_\perp$, except at very low n_e. The phase velocity of the wave parallel to **B** is given by

$$\frac{\omega^2}{k_\parallel^2} = \frac{v_A^2(1 - \Omega_i^2)}{(1 + k_\perp^2 c^2/\omega_{pe}^2)} \tag{12.14}$$

and the ratio of group velocity components is

$$G_v = \frac{\partial\omega/\partial k_\perp}{\partial\omega/\partial k_\parallel} = -\frac{k_\perp\Omega_e\Omega_i}{k_\parallel(1 - \Omega_i{}^2)} \qquad (12.15)$$

In the limit $\Omega_i \to 0$, $G_v \to 0$. At finite Ω_i, $G_v \ll 1$ for laboratory plasma conditions, even when $\Omega_i \to 1$. Consequently, the torsional wave remains strongly guided by magnetic fields at either high or low $k_\perp$ and at frequencies up to the ion cyclotron frequency.

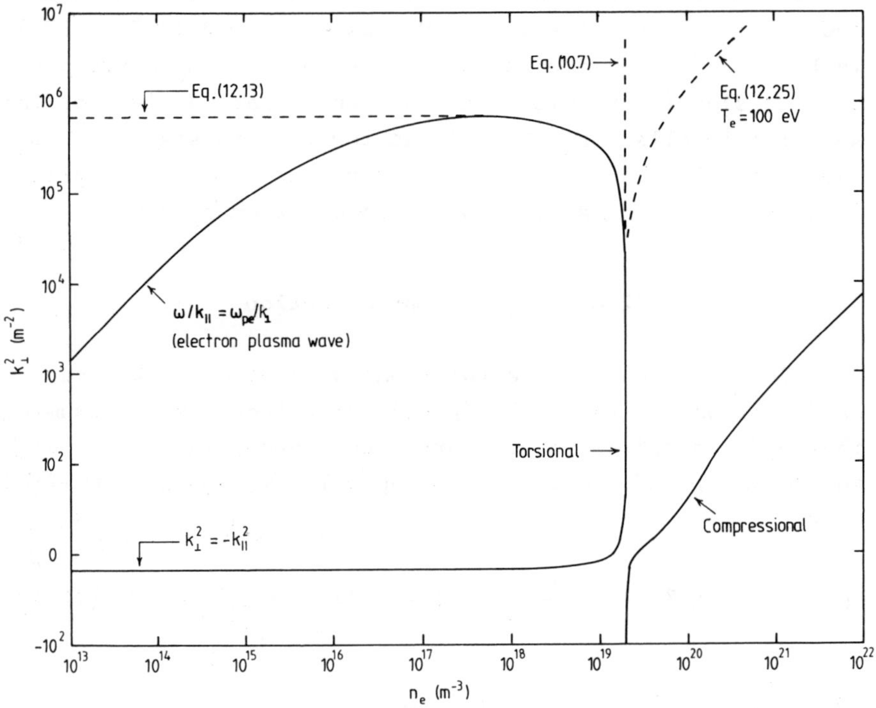

Fig. 12.2 $k_\perp{}^2$ vs n_e for $k_\parallel = 4.0\ m^{-1}$, $\Omega_i = 0.2$ and a hydrogen plasma. The solid curves are solutions of (12.11).

12.5 Wave propagation parallel or perpendicular to B

When **k** is parallel to **B**, $\theta = 0$, the dispersion relation (12.10) reduces to $P(n^4 - 2Sn^2 + RL) = 0$ with solutions $P = 0$, $n^2 = L$ and $n^2 = R$. The solution $P = 0$ corresponds to an oscillation of fixed frequency $\omega^2 = \omega_{pe}{}^2 + \omega_{pi}{}^2$. The solutions $n^2 = L$ and $n^2 = R$ are known as the left and right hand circularly polarised waves respectively. From (12.8), with $n_x = 0$ (when $\theta = 0$),

$$iE_x/E_y = (n^2 - S)/D = D/(n^2 - S)$$

so $iE_x/E_y = \pm 1$. From the definitions of S and D, the solution $n^2 = L$ gives $iE_x/E_y = -1$ (LH) and the solution $n^2 = R$ gives $iE_x/E_y = +1$ (RH). This notation is somewhat misleading since, for propagation angles slightly different to $\theta = 0$, the phase velocity is essentially the same but the polarisation can be quite different. As described in Chapter 9, the polarisation can vary from LH to RH over the plasma cross-section in a bounded plasma and the polarisation can be linear, elliptical or circular depending on the the wave frequency. The LH and RH notation refers strictly only to plane waves with $k_\perp = 0$.

Using the low frequency approximations of Section 12.3, the solution $n^2 = L$ corresponds to a torsional wave with $\omega^2/k_\parallel{}^2 = v_A{}^2(1 - \Omega_i)$ and the solution $n^2 = R$ corresponds to the compressional wave. The fact that these waves are circularly polarised when $\theta = 0$ appears to contradict the results given in Eq.(3.11). However, when $\theta = 0$, both k_x and k_y are zero and the ratios E_y/E_x in Eq.(3.11) are indeterminate. These ratios can only be determined when $\theta = 0$ by including finite frequency terms as described in Chapter 10.

When **k** is perpendicular to **B**, $\theta = \pi/2$ and the solutions of 12.10 are $n^2 = P$ and $n^2 = LR/S$. These solutions are plotted in Fig. 12.3.

12.6 Cold plasma waves

In order to place the Alfven wave modes into perspective, it is useful to consider briefly the other wave types which can propagate in a cold plasma. These are shown in Fig. 12.3 as plots of the phase velocity, v, normalised to the speed of light, c, vs wave frequency for propagation either parallel or perpendicular to **B**. Dispersion relations are normally given as plots of ω vs k, but the plots in Fig. 12.3 are equivalent and give a more direct indication of the phase velocity. When $k_\perp = 0$, equation (12.11) reduces to the form

$$\frac{c^2}{v^2} = 1 - \frac{\omega_{pe}^{\ 2}}{\omega_{ci}\omega_{ce} + \omega^2 \pm \omega(\omega_{ce} + \omega_{ci})} \qquad (12.16)$$

where $\omega_{pe} = (n_e e^2/m_e \epsilon_0)^{\frac{1}{2}}$ is the electron plasma frequency

and where $\omega_{ce} = -\ Be/m_e$ and $\omega_{ci} = +\ Be/m_i$.

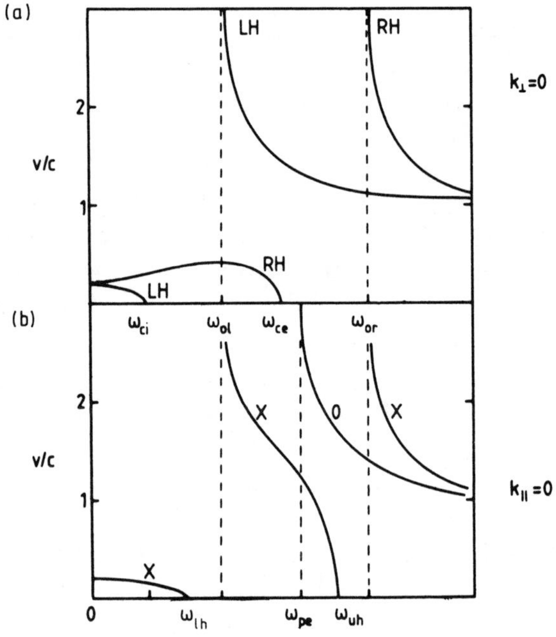

Fig. 12.3 Normalised phase velocity, v/c, vs ω with (a) **k** parallel to **B** (b) **k** perpendicular to **B**

At low wave frequencies, Eq.(12.16) describes (a) the torsional Alfven wave (LH) with a resonance at the ion cyclotron frequency and (b) the compressional Alfven wave (RH) with a resonance at the electron cyclotron frequency. At high wave frequencies, both the LH and RH branches reappear at frequencies above the left and right hand cutoff frequencies given respectively by

$$\omega_{ol} = \left[\omega_{pe}^2 + \frac{(\omega_{ci} - \omega_{ce})^2}{4} \right]^{1/2} + \frac{(\omega_{ce} + \omega_{ci})}{2} \qquad (12.17)$$

$$\omega_{or} = \left[\omega_{pe}^2 + \frac{\omega_{ce}^2}{4} \right]^{1/2} - \frac{\omega_{ce}}{2} \qquad (12.18)$$

If $k_{\parallel} = 0$, so the propagation vector is perpendicular to **B**,

$$\frac{c^2}{v^2} = 1 - \frac{\omega_{pe}^2}{\omega^2} \qquad (12.19)$$

or $$\frac{c^2}{v^2} = 1 - \frac{\omega_{pe}^2}{\left[\omega_{ci}\omega_{ce} + \omega^2 - \dfrac{\omega^2(\omega_{ce} + \omega_{ci})^2}{(\omega_{ci}\omega_{ce} + \omega^2 - \omega_{pe}^2)}\right]} \qquad (12.20)$$

Equation (12.19) describes the ordinary (O) mode and (12.20) describes the extraordinary (X) mode. For the ordinary mode, the wave electric field is parallel to **B** so the wave properties do not depend on B. The same dispersion relation holds even when B = 0. This wave propagates only at frequencies above the plasma frequency, in the far-infrared, or microwave or visible parts of the spectrum. This wave is commonly used, in an interferometer configuration, to measure n_e.

The extraordinary mode, at low wave frequencies, is the compressional Alfven wave which propagates at the Alfven speed regardless of the direction of the propagation vector. The torsional wave does not exist if $k_{\parallel} = 0$. At high wave frequencies, the extraordinary mode has two cutoffs ($v = \infty$), given by (12.17) and (12.18), and two resonances ($v = 0$). The resonance frequencies are known as the lower and upper hybrid frequencies, given respectively by

$$\omega_{lh}^2 = -\omega_{ci}\omega_{ce}\left[\frac{\omega_{pe}^2 - \omega_{ci}\omega_{ce}}{\omega_{pe}^2 + \omega_{ce}^2}\right] \qquad (12.21)$$

$$\omega_{uh}^2 = \omega_{pe}^2 + \omega_{ce}^2 \qquad (12.22)$$

The characteristic frequencies ($f = \omega/2\pi$) are given below for a hydrogen plasma with $n_e = 1 \times 10^{19}$ m^{-3} and B = 1 T.

Characteristic frequencies	**General constants**
f_{ci} = 15.25 MHz	$c = 2.99792 \times 10^8$ m/s
f_{lh} = 465.3 MHz	$e = 1.60219 \times 10^{-19}$ C
f_{ol} = 17.67 GHz	$m_e = 9.1095 \times 10^{-31}$ kg
f_{ce} = 27.99 GHz	$m_p = 1.6726 \times 10^{-27}$ kg
f_{pe} = 28.39 GHz	$k = 1.3807 \times 10^{-23}$ J/K
f_{uh} = 39.87 GHz	$\epsilon_o = 8.8542 \times 10^{-12}$ F/m
f_{or} = 45.65 GHz	$\mu_o = 4\pi \times 10^{-7}$ H/m

12.7 Finite temperature corrections

Finite temperature corrections for the dielectric tensor, determined from kinetic theory, are given by STIX (A8). Wave propagation in a plasma at finite temperature can also be described with the fluid model by retaining ∇p terms in the equation of motion and Ohm's law. This approach represents an important step towards a more realistic description of plasma waves, although it misses out on some of the features, such as Landau damping, of a kinetic theory treatment. The fluid equations, including pressure terms but ignoring the plasma resistivity and finite m_e terms are (SPITZER, A7)

$$\rho \frac{\partial \mathbf{v}}{\partial t} = \mathbf{j} \times \mathbf{B} - \nabla p$$

and

$$\mathbf{E} + \mathbf{v} \times \mathbf{B} = \frac{1}{ne}(\mathbf{j} \times \mathbf{B} - \nabla p_e)$$

where $\quad p = p_e + p_i = C\rho^\gamma \quad$ (C is a constant)

so $\quad \nabla p = a^2 \nabla \rho \quad$ where $\quad a^2 = \gamma p/\rho$

The equation of continuity, $\partial\rho/\partial t = -\rho\, \nabla\cdot\mathbf{v}$, can be linearised, assuming perturbations of the form $\exp[i(\mathbf{k}\cdot\mathbf{r} - \omega t)]$ to give $\delta\rho = (\rho/\omega)\mathbf{k}\cdot\mathbf{v}$ where $\delta\rho$ is a small perturbation in the density ρ. Hence

$$\nabla p = i\mathbf{k}a^2(\delta\rho) = \frac{ia^2\rho}{\omega}\,\mathbf{k}(\mathbf{k}\cdot\mathbf{v})$$

Assuming that $p_e = p_i = p/2$, the above set of equations can be solved, together with Maxwell's equations, to give (BOYD and SANDERSON, A2)

$$\left[\frac{\omega^2}{k^2} - v_A{}^2\cos^2\theta\right]\left[\frac{\omega^4}{k^4} - \frac{\omega^2}{k^2}(a^2+v_A{}^2) + a^2 v_A{}^2\cos^2\theta\right]$$

$$= \Omega_i{}^2 v_A{}^4(\omega^2/k^2 - a^2)\cos^2\theta \qquad (12.23)$$

where $\Omega_i = \omega/\omega_{ci}$. This equation yields the dispersion relations for torsional, compressional and ion-acoustic waves at finite temperature and finite frequency. The effects of kinetic pressure on the dispersion relations are summarised below. The more general form of (12.23), including electron mass, is given in references B7 and B14.

(a) Magnetohydrodynamic waves ($\Omega_i = 0$)

In a cold plasma, there are only two MHD wave types, the torsional and compressional (or fast) Alfven waves. Finite kinetic pressure introduces a third wave type which propagates, when $\mathbf{k}$ is parallel to $\mathbf{B}$, at the acoustic speed, a. If $a < v_A$, this wave is called the slow MHD or ion acoustic wave. It does not propagate across steady magnetic field lines. However, if $a > v_A$, the roles of the slow and fast waves are reversed. The slow wave then propagates at the Alfven speed in a direction parallel to $\mathbf{B}$, and the fast wave propagates at the acoustic speed.

The dispersion relations for the three wave types, at low wave frequencies, can be obtained by setting $\Omega_i = 0$ in (12.23). One solution is then given by $\omega/k = v_A\cos\theta$. This is the dispersion relation for the torsional Alfven wave, which is totally unaffected by the kinetic pressure in this approximation. However, a more careful examination of (12.23) shows that the torsional wave is affected by kinetic pressure at high $k_\perp$, even in the limit $\Omega_i \to 0$. This effect is described below, under the heading "the kinetic Alfven wave". The other two solutions of (12.23), when $\Omega_i = 0$, describe the fast and slow magnetoacoustic waves which satisy the relation

$$\frac{\omega^4}{k^4} - \frac{\omega^2}{k^2}(a^2 + v_A{}^2) + a^2 v_A{}^2 \cos^2\theta = 0 \qquad (12.24)$$

For laboratory conditions, where $a \ll v_A$, the solutions are

$\omega/k = a\cos\theta$ ION ACOUSTIC (or SLOW MHD)

$\omega/k = v_A$ COMPRESSIONAL ALFVEN (or FAST WAVE)

Since $\omega/k_\parallel = a$ for the ion acoustic wave, we see that $\partial\omega/\partial k_\perp = 0$. The group velocity vector is therefore directed exactly parallel to **B**, and the wave is guided by a steady **B** field, as shown in Fig. 6.1.

In space plasmas it is common to encounter conditions where $a > v_A$. The solutions of (12.24) are then not quite as simple as the case $a \ll v_A$, except when $\theta = 0$ or 90°. When $\theta = 0$, the solutions of (12.24) are $\omega/k = a$ (describing the fast wave when $a > v_A$) and $\omega/k = v_A$ (describing the slow wave when $a > v_A$). At $\theta = 90°$, $\omega/k = 0$ or $\omega^2/k^2 = a^2 + v_A{}^2$. The transition from $\theta = 0$ to $\theta = 90°$ is best displayed in a polar phase velocity diagram, as shown in Fig. 12.4.

(b) Propagation near the ion cyclotron frequency

Solutions of (12.23) at finite Ω_i are shown in Fig. 12.4 for several different values of Ω_i and several different values of the ratio a/v_A. The solutions of (12.23) when $\theta = 0$ show that the phase velocity surfaces intersect the $\theta = 0$ axis at

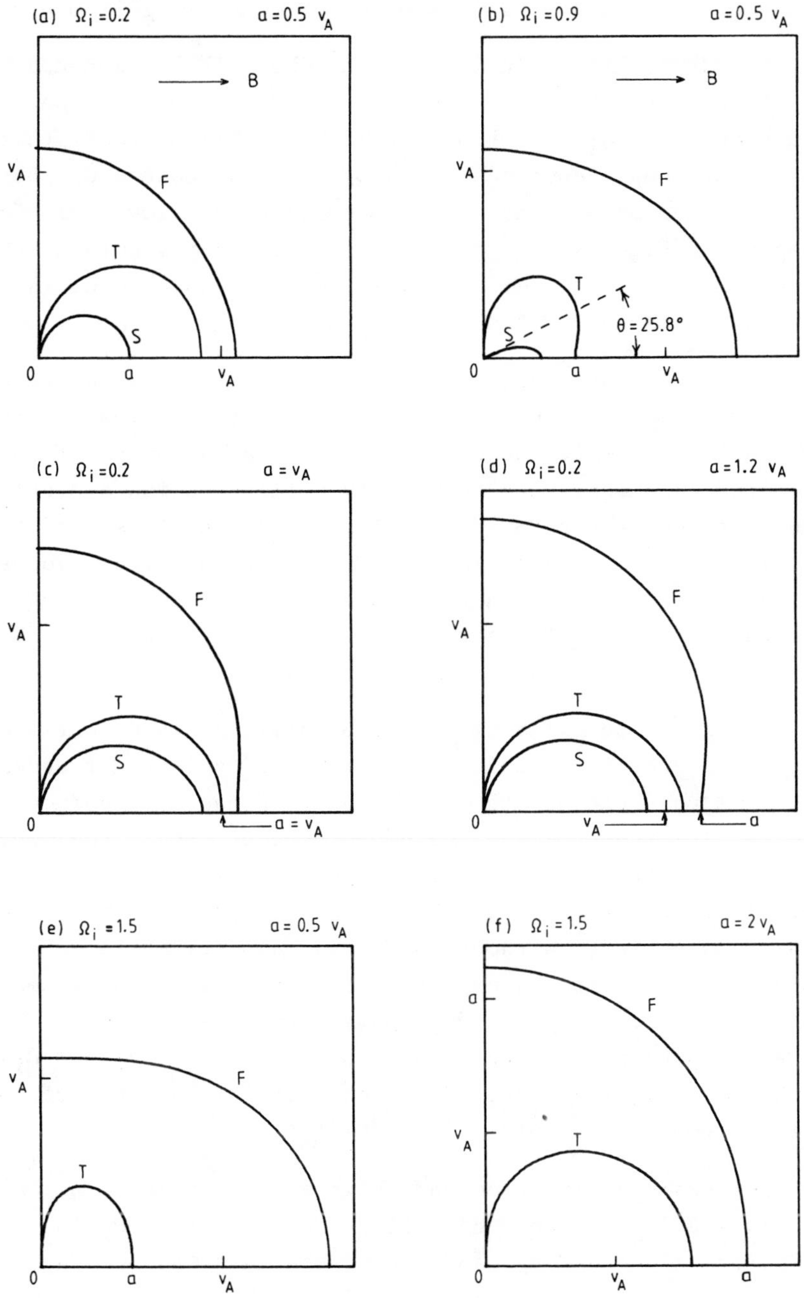

Fig. 12.4 Phase velocity surfaces for fast (F), slow (S) and torsional Alfven (T) waves at finite frequency ($\Omega_i = \omega/\omega_{ci}$) and finite temperature (a = acoustic speed).

$$\omega/k = a \quad \text{or} \quad \omega/k = v_A\,[1 + \Omega_i]^{1/2} \quad \text{or} \quad \omega/k = v_A\,[1 - \Omega_i]^{1/2}$$

The third solution indicates that there is a resonance in one of the wave types at $\Omega_i = 1$ and that this wave type is evanescent when $\Omega_i > 1$. From Eq.(12.23) we find that there is an additional resonance, with $\omega/k = 0$, at $\cos\theta = \Omega_i$. Both of these resonances are associated with the *slow* wave. For example, in Fig. 12.4b, the slow wave propagates only in the range $0 < \theta < 25.8°$. At $\Omega_i = 0.2$, the slow wave propagates only in the range $0 < \theta < 78.5°$.

The three wave types which propagate at low wave frequencies are labelled in Fig. 12.4 as fast (F), slow (S) and torsional (T). At frequencies near the ion cyclotron frequency, the torsional wave is usually called the ion cyclotron wave, but the labels F, S and T are used in all six of the diagrams in Fig. 12.4 to show how each wave type evolves from one diagram to the next. These labels are consistent with the magnitude of the respective phase velocities, but close inspection reveals some surprises.

The first surprise is that the torsional wave continues to propagate at frequencies above the ion cyclotron frequency. In a cold plasma, the torsional wave has a resonance at $\Omega_i = 1$. The reader might therefore suspect that the wave labelled "T" in Fig. 12.4b or e should be labelled "S", but there is no mistake. The three wave types all evolve, as Ω_i or a/v_A is varied, without the phase velocity surfaces ever crossing. The surfaces expand or contract as these parameters are varied, but never pass through each other. The nearest approach to a crossing occurs when $a = v_A(1 + \Omega_i)^{1/2}$ or when $a = v_A(1 - \Omega_i)^{1/2}$. For these conditions, two of the surfaces touch at a point along the $\theta = 0$ axis.

For example, if we compare Figs. 12.4a, b and e, all with $a = 0.5v_A$, then the effect of varying Ω_i is the following. When $a < v_A$, and $\theta = 0$, the fast wave propagates at a phase velocity $\omega/k = v_A(1 + \Omega_i)^{1/2}$. The torsional wave propagates, at low Ω_i and $\theta = 0$, at a phase velocity $\omega/k = v_A(1-\Omega_i)^{1/2}$. As Ω_i increases, the phase velocity of the torsional wave decreases to a but does not decrease below a.

As Ω_i is increased further, the phase velocity of the torsional wave remains equal to a but the velocity of the slow wave decreases to zero as $\Omega_i \to 1$. Only the fast and torsional waves propagate when $\Omega_i > 1$.

Similarly, if we compare Figs. 12.4a,c and d, all with $\Omega_i = 0.2$, then the effect, at $\theta = 0$, of increasing a/v_A is the following. When $a/v_A \ll 1$, the slow wave propagates at a phase velocity $\omega/k = a$ and the torsional wave propagates at a phase velocity $\omega/k = v_A(1 - \Omega_i)^{1/2}$. As a/v_A increases, the phase velocity of the slow wave increases to $v_A(1-\Omega_i)^{1/2}$. With a further increase in a/v_A, the phase velocity of the slow wave remains fixed at $v_A(1 - \Omega_i)^{1/2}$ and the phase velocity of the torsional wave increases to $v_A(1 + \Omega_i)^{1/2}$. At even higher values of a/v_A, the phase velocity of the torsional wave remains fixed at $v_A(1 + \Omega_i)^{1/2}$ and the fast wave propagates with $\omega/k = a$.

The F, T and S wave types cannot be identified unambiguously simply from the values of ω/k at $\theta = 0$. Correct identification requires an examination of the phase surfaces at an additional or alternative value of θ, say $\theta = 45°$, to establish the fact that the phase surfaces never cross.

The slow wave has been studied extensively in very low density plasmas, but not in tokamak plasmas. In laboratory plasmas, the phase velocity of the slow wave, in a direction parallel to **B**, is equal to the acoustic speed. When the electron and ion temperatures are comparable, as in tokamak plasmas, the acoustic speed is approximately equal to the ion thermal speed $(3kT_i/m_i)^{1/2}$. As a result, the slow wave is Landau damped by the ions within a few parallel wavelengths. In very low density plasmas, where $T_e \gg T_i$, Landau damping is not as severe, and the wave is easily generated and detected. Alternatively, if the plasma is sufficiently collisional (high n_e and low T_e), then the slow wave can be observed even when $T_e = T_i$, as in Fig. 6.1, since Landau damping is not effective in a collisional plasma. A close relative of the slow wave is the drift wave which arises as a result of pressure gradients in the plasma. At high $k_\perp$, the drift and slow waves are the same wave type.

(c) The kinetic Alfven wave

As described above, the term on the right side of (12.23) remains important, even in the limit $\Omega_i \to 0$, at large values of $k_\perp$. This term gives a finite Larmor radius (FLR) correction, of order $k_\perp \rho_i$, to the ideal MHD results, where ρ_i is the ion Larmor radius (ie the gyroradius). The correction is therefore significant when describing short perpendicular wavelength torsional Alfven waves, but is usually negligible for the low $k_\perp$ compressional wave.

Eq. (12.23) is of the form $Ak_\perp{}^4 + Bk_\perp{}^2 + C = 0$ where $k_\perp = k \sin \theta$ and $k_\parallel = k \cos \theta$. Approximate solutions for large $k_\perp$ can be obtained by ignoring C. If we also assume that $a \ll v_A$ and that $k_\perp \gg k_\parallel$ then (12.23) reduces to

$$\frac{\omega^2}{k_\parallel{}^2 v_A{}^2} = 1 - \Omega_i{}^2 \left[1 - \frac{a^2 k_\perp{}^2}{\omega^2}\right] \tag{12.25}$$

which is the fluid model dispersion relation for the kinetic Alfven wave. The kinetic theory version (A6, E11, E21), at low wave frequencies, is

$$\frac{\omega^2}{k_\parallel{}^2 v_A{}^2} = 1 + k_\perp{}^2 \rho_i{}^2 \left[\frac{3}{4} + \frac{T_e}{T_i}\right] \tag{12.26}$$

If we take $T_e = T_i$, $\gamma = 5/3$ and $\rho_i{}^2 = a^2/\omega_{ci}{}^2$, then the fluid and kinetic dispersion relations are the same except for a slight difference in the factor multiplying $k_\perp{}^2 \rho_i{}^2$. Eq. (12.25) is plotted in Fig. 12.2. As shown in the figure, the kinetic Alfven wave propagates, with short perpendicular wavelength, on the high density side of the Alfven resonance layer.

It is widely claimed in relation to the Alfven wave heating scheme that the kinetic Alfven wave propagates across field lines. Let us examine this claim. From (12.25)

$$\omega^2 = \frac{k_\parallel{}^2 v_A{}^2 (1 + k_\perp{}^2 a^2/\omega_{ci}{}^2)}{(1 + k_\parallel{}^2 v_A{}^2/\omega_{ci}{}^2)}$$

so

$$G_v = \frac{\partial\omega/\partial k_\perp}{\partial\omega/\partial k_\parallel} = \frac{\omega^2\,\omega_{ci}}{k_\parallel{}^3 k_\perp a^2 v_A{}^2} \sim \frac{1}{k_\perp k_\parallel \rho_i{}^2}$$

since $\omega \sim k_{\parallel} v_A$. In an Alfven resonance layer, $k_\perp \gg k_\parallel$ and $k_\perp \rho_i \leq 1$, so $G_v \ll 1$. Consequently, the kinetic Alfven wave propagates almost exactly along field lines. Most of the wave energy will therefore be dissipated near the resonance layer. The main damping mechanism is Laudau damping by electrons since the electron thermal speed is comparable with the Alfven speed in high temperature plasmas.

(d) Ion Bernstein waves

In a single ion species plasma, and at frequencies in the range $\omega_{ci} \leq \omega \leq 10\omega_{ci}$, the only cold plasma wave of interest for ICRF heating is the compressional Alfven wave. The electron plasma wave can also propagate in this frequency range, but only at very low electron densities (eg at the extreme edge of the plasma). The hot plasma dispersion relation (STIX, A8) admits additional wave types, one of the most important being the ion Bernstein wave. This wave has no resemblance to any of the cold plasma wave types. It propagates in frequency bands above the fundamental and the harmonics of the ion cyclotron frequency, as shown in Fig. 12.5, with a short perpendicular wavelength at frequencies close to the fundamental and the harmonics. The ion Bernstein wave is electrostatic, with **E** parallel to **k**. The warm plasma dispersion relation for such waves has the form (A8)

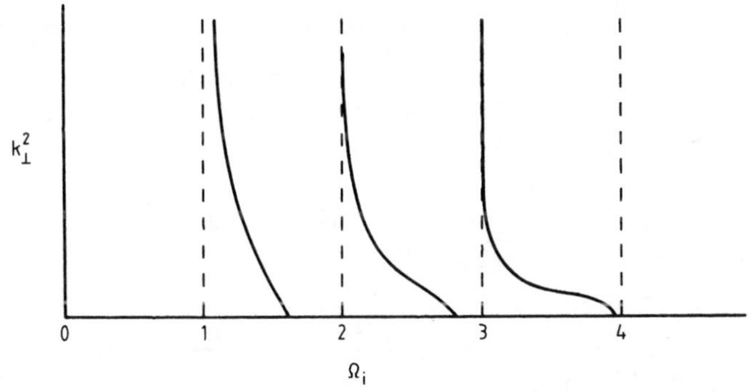

Fig. 12.5 Ion Bernstein dispersion relation when $k_\parallel = 0$.

$$k_\perp^2 K_{xx} + k_\parallel^2 K_{zz} = 0 \quad (12.27)$$

Solutions of Eq.(12.27) are given in Refs. (A8,I10,I11,J9). At frequencies near the fundamental and cyclotron harmonics, the phase velocity in a direction perpendicular to B is comparable with the ion acoustic speed and is a function of the ion temperature. The group velocity vector has a strong component perpendicular to the steady B field, allowing good penetration of the wave to the plasma interior.

The ion Bernstein wave is involved in mode conversion of the fast wave at the ion–ion hybrid resonance layer in a two ion species plasma. In the cold plasma model, $k_\perp^2 \to \infty$ at the resonance layer. In a warm plasma model, a small $\lambda_\perp$ ion Bernstein wave is generated at the resonance layer. This wave therefore plays a role similar to the kinetic Alfven wave at an Alfven resonance layer in resolving the physically unrealistic cold plasma result that $\lambda_\perp = 0$ at the resonance layer.

Some nice experimental results on direct excitation of the ion Bernstein wave have been obtained in the small ACT–1 device (J9, J10), and significant wave heating by this means has been observed in several large tokamaks (eg, J11). The wave is usually excited with a loop, carrying current parallel to the steady B field, and electrostatically shielded in the poloidal direction to avoid excitation of the fast wave. This type of antenna couples to the Bernstein wave via the E_z wave field component. In ion Bernstein wave heating experiments, the antenna is located on the low field side of the torus, and the wave propagates towards a cyclotron harmonic layer at the centre of the plasma where it is cyclotron damped.

An interesting feature of this scheme is that it has broken a conventional taboo in ICRF experiments. All fast wave antennas are electrostatically shielded to eliminate the E_z component. The shield also doubles as a particle shield, and this function may be more significant in some experiments. At frequencies below the ion cyclotron frequency, the E_z component couples to high $k_\perp$ torsional or ion–ion hybrid waves, leading to edge heating of the plasma (G1).

13. PLASMAS WITH TWO ION SPECIES

One of the most successful RF heating schemes is the minority heating scheme, which involves Alfven wave propagation in a plasma with two ion species. The gas mixture chosen normally contains only a small percentage of the minority species and the wave frequency is chosen to be near the ion cyclotron frequency of this species. As a result, this species is selectively heated and then transfers energy to the other species and to electrons by Coulomb collisions. Most experiments are conducted in deuterium containing up to 5% hydrogen or ^{3}He as the minority species. The ^{3}He minority experiments model minority deuterium in tritium in the sense that the fundamental cyclotron frequency of the minority does not coincide with the second harmonic of the majority as it does in a hydrogen - deuterium mixture.

The scheme originated from attempts to heat pure deuterium at a wave frequency $\omega = 2\omega_{ci}$. Heating at ω_{ci} is not expected in a single species plasma since the compressional wave, which carries energy from an external antenna to the plasma interior, has the wrong polarisation at ω_{ci}. At $\omega = \omega_{ci}$, it is right hand circularly polarised. At $\omega > \omega_{ci}$ it is right hand elliptically polarised, which is equivalent to having a strong right hand circularly polarised component and a weaker left hand circularly polarised component. By operating at $2\omega_{ci}$, the ions will be heated by the left hand component, provided there is a gradient in the wave electric field so that the acceleration on one side of an ion orbit is greater than the deceleration on the other side. Wave heating can therefore be expected if $k_\perp \neq 0$.

The $\omega = 2\omega_{ci}$ experiments turned out to be more successful than expected. The success was attributed to a small fraction of impurity hydrogen in the deuterium plasma. The hydrogen ions were being heated at their fundamental cyclotron frequency. This effect in turn was due to an increase in the left hand circularly polarised component of the wave fields caused by the proximity of the ion-ion hybrid resonance. This resonance, described below, occurs at a frequency between the cyclotron frequencies of the two species and is close to the cyclotron frequency of the minority species.

In tokamak geometry, the toroidal field varies over the plasma cross-section inversely as the major radius, R. The cyclotron resonance layers are vertical surfaces in the cross-section and the ion-ion hybrid resonance layer forms another nearly vertical surface between the two cyclotron layers. The ion-ion hybrid surface moves closer to the ω_{ch} surface as the hydrogen concentration is reduced.

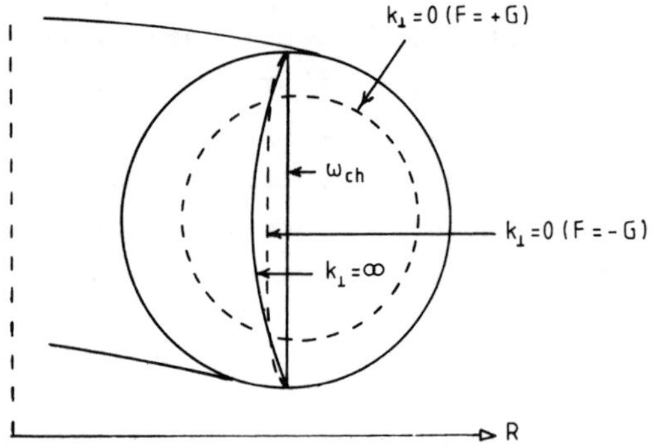

Fig. 13.1 Resonance ($k_\perp = \infty$) and cutoff ($k_\perp = 0$) surfaces in a hydrogen-deuterium plasma for a parabolic density variation, with ~5% hydrogen and ~95% deuterium, when the wave frequency matches the hydrogen cyclotron frequency at the plasma centre.

Research work on the ICRF heating schemes is being pursued vigorously. Almost all large tokamak devices employ ICRF heating at power levels above 1MW. Debate continues on the best location of the antenna (high field or low field side excitation), the best orientation of the antenna (parallel or perpendicular to **B**), the best gas mix, the relative merits of heating at ω_{ch} or its harmonics or at the ion-ion hybrid resonance frequency, on the sources of impurity generation observed in all experiments and on the various theoretical models (ray tracing or full wave calculations).

In this Chapter we consider the most elementary properties of wave propagation in a two-ion species plasma, by examining the cold plasma dispersion relation for a current - free plasma. Some general results are obtained for ions of arbitrary q/m and specific calculations are given for a H - D plasma with $n_h/n_d = 0.2$.

13.1 Cold plasma dispersion relations

The dispersion relation for a plasma with two ion species is given by equation (12.11). This relation includes finite m_e and displacement current. It is easily solved for $k_\parallel$ given $k_\perp$, or vice versa. The essential features can be seen more clearly if we make the approximations $\omega \ll \omega_{ce}$ and $v_A \ll c$ in the expressions for P,R,L,S and D.

Let n_1 and n_2 be the number densities of the two ion species and let Z_1 and Z_2 be their charge numbers. Then $n_e = Z_1 n_1 + Z_2 n_2$,

$$\frac{\omega_{pe}^2}{\omega_{ce}} = -\frac{(1+x)c^2\omega_{c1}}{v_1^2} \qquad \frac{\omega_{p1}^2}{\omega_{c1}} = \frac{c^2\omega_{c1}}{v_1^2} \qquad \frac{\omega_{p2}^2}{\omega_{c2}} = x\,\frac{c^2\omega_{c1}}{v_1^2}$$

$$\omega_{c1} = \frac{BZ_1e}{m_1} \qquad \omega_{c2} = \frac{BZ_2e}{m_2} \qquad x = \frac{Z_2n_2}{Z_1n_1} \qquad v_1^2 = \frac{B^2}{\mu_0 n_1 m_1}$$

Now let $\Omega_1 = \omega/\omega_{c1}$, $\Omega_2 = \omega/\omega_{c2}$ and $\Omega_e = -\,\omega/\omega_{ce}$.
When $\omega \ll \omega_{ce}$ and $v_1 \ll c$,

$$P = -\frac{(1+x)c^2}{\Omega_e\Omega_1 v_1^2} \tag{13.1}$$

$$R = \frac{c^2}{\Omega_1 v_1^2}\left[(1+x) - \frac{1}{1+\Omega_1} - \frac{x}{1+\Omega_2}\right] \tag{13.2}$$

$$L = \frac{c^2}{\Omega_1 v_1^2}\left[-(1+x) + \frac{1}{1-\Omega_1} + \frac{x}{1-\Omega_2}\right] \tag{13.3}$$

$$S = \frac{R+L}{2} = \frac{c^2}{\Omega_1 v_1^2}\left[\frac{\Omega_1}{1-\Omega_1^2} + \frac{x\,\Omega_2}{1-\Omega_2^2}\right] \tag{13.4}$$

$$D = \frac{R-L}{2} = \frac{c^2}{\Omega_1 v_1^2}\left[(1+x) - \frac{1}{1-\Omega_1^2} - \frac{x}{1-\Omega_2^2}\right] \tag{13.5}$$

If $\Omega_e \ll 1$, then $S \ll P$ and $LR \ll SP$ except in the immediate vicinity of the cyclotron or the ion–ion hybrid resonance frequencies where $S \to \infty$ or $S \to 0$. Assuming that ω is not very close to a resonance and that $Sk_\perp^4 \ll Pk_\parallel^4$, then (12.11) reduces, with the aid of the identity $LR = S^2 - D^2$, to

$$\boxed{k_\perp^2 = \frac{F^2 - G^2}{F}} \qquad (13.6)$$

where $\boxed{F = A - k_\parallel^2}$ $\boxed{A = \omega^2 S/c^2}$ $\boxed{G = -\omega^2 D/c^2}$

Alternatively, if the term Sn_x^4 is ignored in (12.9), and if $\Omega_e \ll 1$ then (13.6) has the equivalent form

$$\boxed{n_\perp^2 = \frac{(S - n_\parallel^2)^2 - D^2}{S - n_\parallel^2}} \qquad (13.7)$$

The $n_\perp = \infty$ or $k_\perp = \infty$ ion–ion hybrid resonance surface is defined by $n_\parallel^2 = S$ or by $F = 0$. The $n_\perp = 0$ or $k_\perp = 0$ cutoff surfaces are defined by $n_\parallel^2 = L$ and $n_\parallel^2 = R$, or equivalently by $F = -G$ and $F = +G$. If the term $Sk_\perp^4$ is retained in (12.11), then $k_\perp$ is large at the ion – ion hybrid resonance but it is not infinite.

Equation (13.6) is quadratic in $k_\parallel^2$ with solutions

$$k_\parallel^2 = A - k_\perp^2/2 \pm [G^2 + k_\perp^4/4]^{1/2} \qquad (13.8)$$

Results for a single ion species are recovered if we let $n_1 = 0$ or $n_2 = 0$, in which case the ± signs give torsional and compressional wave modes. These modes are still identifiable in a two–ion species plasma. However, the torsional wave has two $k_\parallel = \infty$ resonances, one at the lower cyclotron frequency and one at the higher cyclotron frequency. At frequencies near the higher cyclotron frequency, this wave is known as the ion–ion hybrid wave.

13.2 Propagation parallel or perpendicular to B

It is rare in laboratory plasmas to observe wave propagation in a direction exactly parallel or perpendicular to B. Normally, Alfven waves are observed at angles, θ, other than 0° or 90°. Nevertheless, it is instructive to examine these special cases. When $k_\perp = 0$ ($\theta = 0°$) Eq.(13.6) gives $F = \pm G$ so $k_\parallel^2 = A \pm G$. When $k_\parallel = 0$ ($\theta = 90°$), Eq.(13.6) gives $k_\perp^2 = (A + G)(A - G)/A$.

For purposes of illustration, consider a H – D plasma with $n_h = n_1$, $n_d = n_2$, $Z_1 = Z_2 = 1$, $x = n_d/n_h$ and $\Omega_2 = 2\Omega_1$ since $m_d = 2m_h$. Then

$$A + G = \frac{\omega^2 L}{c^2} = \frac{y(a - b\Omega)}{(1 - \Omega)(1 - 2\Omega)}$$

$$A - G = \frac{\omega^2 R}{c^2} = \frac{y(a + b\Omega)}{(1 + \Omega)(1 + 2\Omega)}$$

$$\frac{(A + G)(A - G)}{A} = \frac{y(a^2 - b^2\Omega^2)}{[a - (b + 2)\Omega^2]}$$

where $\Omega = \Omega_1 = \omega/\omega_{ch}$, $a = 1 + 2x$, $b = 2 + 2x$ and

$$y = \frac{\omega^2}{v_1{}^2} = \left[\frac{\mu_o e^2}{m_h}\right] n_h\,\Omega^2 = 1.92 \times 10^{-17} n_h\,\Omega^2$$

The dispersion relations for propagation perpendicular or parallel to **B** are, therefore,

$$\frac{\omega^2}{k_\perp{}^2 v_A{}^2} = \frac{a[a - (b + 2)\Omega^2]}{a^2 - b^2\Omega^2} \qquad (13.9)$$

and

$$\frac{\omega^2}{k_\parallel{}^2 v_A{}^2} = \frac{a(1 - \Omega)(1 - 2\Omega)}{(a - b\Omega)} \qquad (13.10)$$

or

$$\frac{\omega^2}{k_\parallel{}^2 v_A{}^2} = \frac{a(1 + \Omega)(1 + 2\Omega)}{(a + b\Omega)} \qquad (13.11)$$

where $\omega^2/v_A{}^2 = ay$ and $v_A{}^2 = \dfrac{B^2}{\mu_o(n_h m_h + n_d m_d)}$

These relations are shown in Fig. 13.2 for $n_h/n_d = 0.2$, $a = 11$, $b = 12$. Only one wave type, the compressional Alfven wave, can propagate perpendicular to **B**. This wave has a $k_\perp = \infty$ resonance at the ion–ion hybrid resonance frequency $\Omega = \Omega_r$ given, from (13.9), by

$$\Omega_r{}^2 = \frac{a}{b+2} = \frac{1 + 2n_d/n_h}{4 + 2n_d/n_h} \tag{13.12}$$

It also has a $k_\perp = 0$ cutoff, at $\Omega = \Omega_c$, given by

$$\Omega_c = \frac{a}{b} = \frac{1 + 2n_d/n_h}{2 + 2n_d/n_h} \tag{13.13}$$

Both Ω_r and $\Omega_c \to 1$ as $n_h/n_d \to 0$. In other words, the resonance and cutoff frequences are close to the cyclotron frequency of the *minority* species.

The dispersion relations for propagation parallel to **B** are shown in Fig. 13.2b. At frequencies below the deuterium cyclotron frequency, $\Omega < 0.5$, it is easy to identify the two wave types as torsional and compressional waves. The

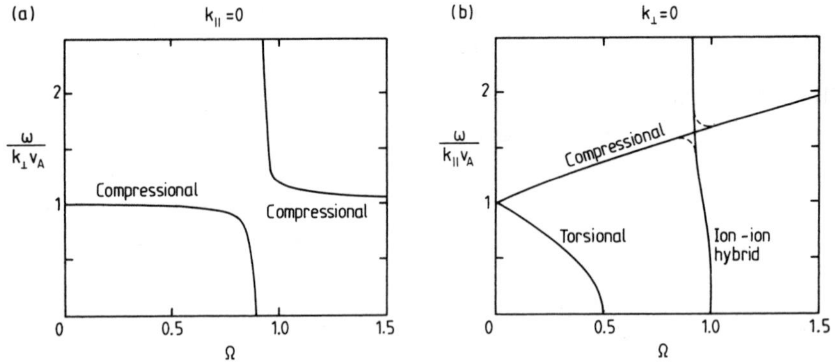

Fig. 13.2 Dispersion relations for propagation (a) perpendicular and (b) parallel to **B** in a hydrogen – deuterium plasma with $n_h/n_d = 0.2$, $\Omega_r = 0.886$, $\Omega_c = 0.917$, $\Omega = \omega/\omega_{ch}$.

compressional wave continues to propagate, parallel to **B**, at frequencies well above the hydrogen cyclotron frequency. The torsional wave has a resonance at the deuterium cyclotron frequency. It re-appears, at frequencies just below $\Omega = 1$, as the ion-ion hybrid wave. The ion-ion hybrid wave propagates in the frequency range $\Omega_c < \Omega < 1$.

When all terms are retained in equation (12.11), the compressional and ion-ion hybrid dispersion relations do not intersect. Rather, the dispersion relations merge without intersecting as indicated by the dashed lines in Fig. 13.2b. The ion-ion hybrid wave has physical properties intermediate between those of the torsional and compressional Alfven wave types. Some of these properties, and the nature of the ion-ion hybrid resonance, are described below.

13.3 Phase and group velocity surfaces

Phase and group velocity surfaces, with $n_h/n_d = 0.2$, are shown in Fig. 13.3. These surfaces were computed using Eqs. (5.3) and (5.4), retaining all terms in the dispersion relation, Eq. (12.11). Surfaces at $\Omega < 0.5$ are not shown, since they are very similar to those for a single ion species plasma (Fig. 5.3). The phase velocity plots show that the compressional wave, at $\Omega = 0.6$, evolves into an ion-ion hybrid wave as Ω increases, having a cyclotron resonance at $\Omega = 1$. This evolution is shown by the dashed curve in Fig. 13.2b. The ion-ion hybrid wave, at $\Omega = 0.9$, evolves into a compressional wave as Ω increases. Only one wave type (the compressional wave) propagates in the frequency range $0.5 < \Omega < \Omega_c$ or when $\Omega > 1$.

The group velocity plots reveal that the ion-ion hybrid wave is essentially isotropic near the cutoff eg. at $\Omega = 0.935$, but is strongly anisotropic at $\Omega \sim 1$. Wave energy is magnetically guided along the steady field lines, as observed experimentally (Fig. 6.5). The compressional wave is slightly anisotropic at frequencies well removed from the ion-ion hybrid resonance. At frequencies between the hybrid resonance and ion cyclotron frequencies, the compressional wave is also guided by the steady magnetic field.

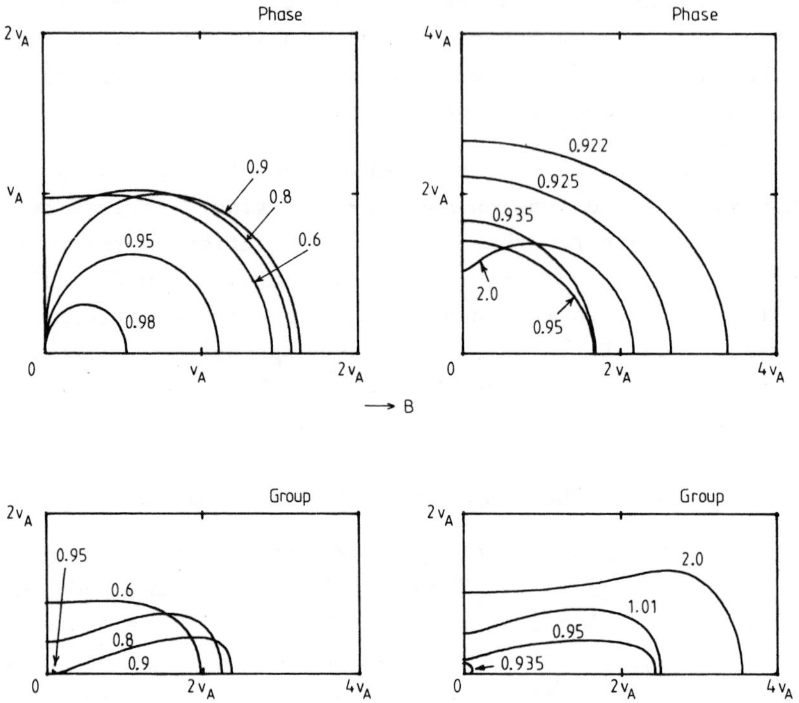

Fig. 13.3 Phase and group velocity surfaces in a hydrogen-deuterium plasma, with $n_h/n_d = 0.2$. The surfaces are plotted in the first quadrant only, for selected values of $\Omega = \omega/\omega_{ch}$.

13.4 Propagation at fixed $\mathbf{k}_\perp$ or fixed $\mathbf{k}_\parallel$

The phase velocity of Alfven and ion - ion hybrid waves in a direction parallel to **B** is shown in Fig. 13.2b for the special case $k_\perp = 0$. Fig. 13.4 shows similar plots for $k_\perp = 10\ m^{-1}$ and $k_\perp = 50\ m^{-1}$. The phase velocity of the torsional and ion - ion hybrid waves in a direction parallel to **B** is almost independent of the value of $k_\perp$. However, the compressional wave has a $k_\parallel = 0$ cutoff frequency proportional to $k_\perp$. At high values of $k_\perp$, the compressional wave propagates only at frequencies well above the cyclotron frequencies.

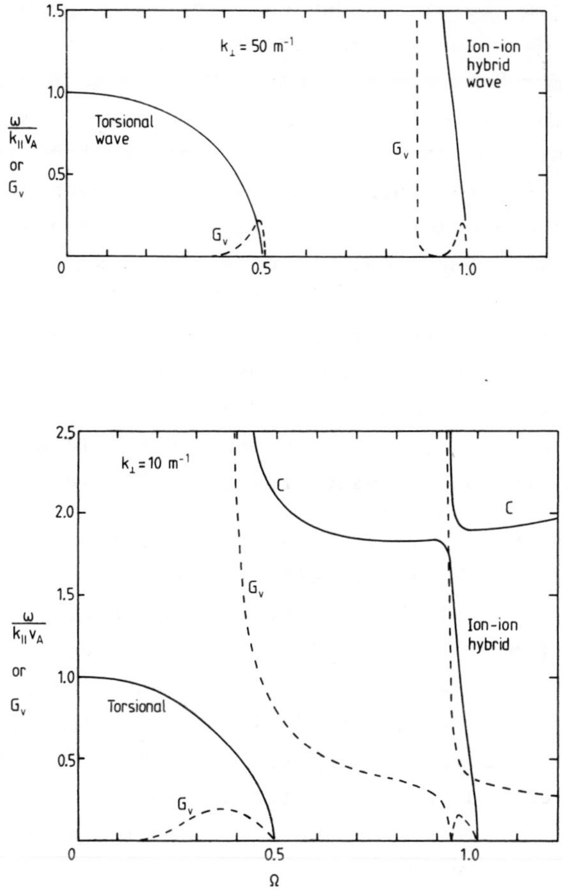

Fig. 13.4 $\omega/k_{\parallel}v_A$ (solid curves) and G_v (dashed curves) vs Ω for a hydrogen – deuterium plasma with $n_e = 2 \times 10^{19}$ m^{-3}, $n_h/n_d = 0.2$, and $k_{\perp} = 10$m^{-1} or 50m^{-1}. C denotes compressional wave. $\Omega = \omega/\omega_{ch}$, $G_v = (\partial\omega/\partial k_{\perp})/(\partial\omega/\partial k_{\parallel})$.

The results shown in Fig. 13.4 are given for fixed values of $k_{\perp}$ to highlight the behaviour of torsional and ion – ion hybrid waves. These waves are guided along helical field lines and do not return to the antenna in tokamak devices, so there is no $k_{\parallel}$ selection (ie no standing waves are formed). There is also no $k_{\perp}$ selection since these waves do not reflect off the boundary walls. Any given antenna will generate a continuous spectrum of $k_{\perp}$ components. The values of $k_{\perp}$ used in Fig. 13.4 are chosen to be typical.

When describing the behaviour of compressional wave eigenmodes in a toroidal device, it is conventional to assume that $k_{\parallel}$ is determined by the periodic boundary condition $n\lambda_{\parallel} = 2\pi R$. In this case, $k_{\perp}^2 \to \infty$ as $F \to 0$ according to equation (13.6). The surface where $k_{\perp}^2 = \infty$ is called the ion–ion hybrid resonance layer. This surface is plotted in Figure 13.1, together with the $k_{\perp} = 0$ resonance layers. Similar plots are common in the literature and are obtained from Eq. (13.6) using local values of n_e, B and $k_{\parallel} = n/R$ for integer $n \sim 3$. The location of the cutoff and resonance surfaces is not very sensitive to $k_{\parallel}$ or n_e, unless n_e is very small or $n_{\parallel} > 6$. It should be noted, however, that the plasma current is assumed to be zero in this Chapter, so $k_{\parallel} = k_z$. The effect of plasma current on the location of the resonance layers is discussed briefly in Section 13.7.

As shown in Fig. 13.1, the hybrid resonance surface is located close to a $k_{\perp} = 0$ cutoff for the compressional wave. As a result, some of the wave energy approaching the resonance surface from the low field (large R) side of the torus will be reflected when it reaches the cutoff. This effect does not occur if the wave is incident from the low field side. In the *minority heating scheme*, the external antenna is normally located on the low field side of the torus and generates a compressional wave which is focussed by the density gradient towards the centre of the plasma. A significant fraction of the wave energy is absorbed by cyclotron damping before reaching the cutoff layer.

In the *wave conversion heating scheme*, the antenna is normally located on the high field side of the torus and the minority concentration is relatively high, typically ~20%. The compressional wave generated by the antenna undergoes mode conversion to an ion Bernstein wave at the hybrid resonance layer. Wave heating is due to a combination of (a) Landau damping of the Bernstein wave, which heats the electrons, and (b) ion cyclotron damping of the Bernstein wave, which heats the minority ions.

The $k_{\perp} = 0$ cutoff surface near the plasma edge (the $F = + G$ surface in Fig. 13.1) must also be crossed by the compressional wave if it is generated by an antenna at the plasma edge. The wave is evanescent, with $k_\perp^2 < 0$ between the plasma edge and the cutoff surface. However, $k_\perp^2$ remains small so the wave is only weakly evanescent and the wave fields are not strongly attenuated between the plasma edge and the cutoff surface. Although the compressional wave is evanescent near the plasma edge, there exists a range of frequencies, described above, where the torsional and ion-ion hybrid waves are not. These waves can propagate along steady field lines at the plasma edge and may therefore contribute to edge heating (G2).

13.5 Wave field polarisation

Part of the success of the minority heating scheme is due to the favourable wave field polarisation near the minority cyclotron frequency. The polarisation can be determined from equation (12.8),

$$\frac{iE_x}{E_y} = \frac{c^2}{\omega^2 D}\left[k_\perp^2 + k_\parallel^2 - \frac{\omega^2 S}{c^2}\right] \tag{13.14}$$

This relation was derived for the special case $k_y = 0$ where

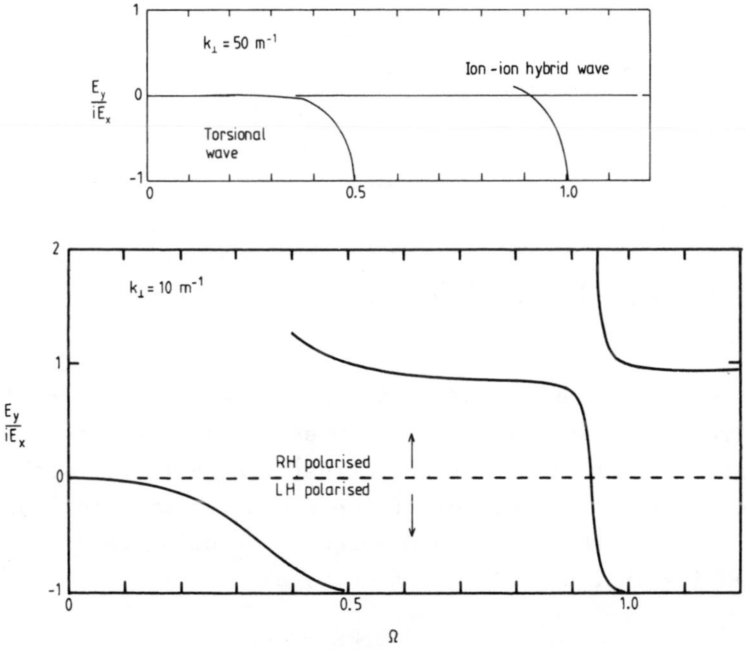

Fig. 13.5 Ratio E_y / iE_x vs Ω when $k_y = 0$ for the same conditions as those in Fig. 13.4

$k_\perp{}^2 = k_x{}^2$. The case where $k_y \neq 0$ is more complicated. For a hydrogen–deuterium plasma and for $\omega \ll \omega_{ce}$, S and D are given by equations (13.4) and (13.5). The ratio E_y/iE_x is shown in Fig. 13.5 for the same conditions as those in Fig. 13.4. Using the method of Sec. 10.3, one can show that $v_y/iv_x = -1$ for ion gyromotion and hence the wave fields are LH polarised when $E_y/iE_x < 0$ and RH polarised when $E_y/iE_x > 0$.

13.6 The ion–ion hybrid resonance

The ion–ion hybrid frequency is a natural frequency of oscillation of the ions in a two ion species plasma, where ions of different q/m rotate 180° out of phase. This effect differs from cyclotron orbital motion in that coupling between the two ion species arises from an induced space charge field. This field depends on the density of the two species and hence the oscillation frequency varies as n_1/n_2 is varied. Suppose that we have a mixture of ions confined by a B field in the z direction and that the two separate fluids behave cooperatively to produce a transverse field E_x as shown below.

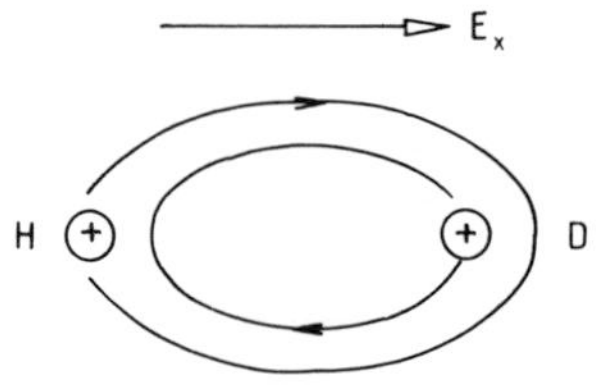

Fig. 13.6 Ion–ion hybrid guiding centre motion.

It will be seen from (13.14) that $E_y = 0$ at the ion–ion hybrid resonance when $k_y = 0$, so an essential feature of the resonance is that it is an electrostatic effect, with **E** parallel to **k**. The behavior of the plasma at the resonance frequency can therefore be deduced by assuming that the E_x field varies in the x direction only, as

$$E_x = E_o e^{i(kx - \omega t)}$$

Then $\nabla \cdot \mathbf{E} = ikE_x = \dfrac{\rho}{\epsilon_o} = \dfrac{(q_1\Delta n_1 + q_2\Delta n_2)}{\epsilon_o}$

where ρ is the local charge density and Δn is a small perturbation in the ion number density. It is assumed that $\partial/\partial y = \partial/\partial z = 0$ and that there is no change in the electron number density since the resonance involves only relative displacement of the two ion species (BUCHSBAUM, B5).

Since $$\frac{\partial n}{\partial t} + n\nabla \cdot \mathbf{v} = 0 \qquad \text{for each species}$$

or $$-i\omega\Delta n + nikv_x = 0$$

then $$i\omega\epsilon_o E_x = n_1 q_1 v_{x1} + n_2 q_2 v_{x2}$$

The displacement current $\epsilon_o\partial\mathbf{E}/\partial t$ is therefore equal and opposite to the conduction current $\mathbf{j} = ne\mathbf{v}$, so the total current is zero. The same situation arises in plasma oscillations at the electron plasma frequency ω_{pe}. The velocity $\mathbf{v}$ in the above expressions is a guiding centre drift velocity rather than a single particle velocity. For each ion species the drift velocity $\mathbf{v}$ is related to $\mathbf{E}$ and $\mathbf{B}$ by

$$m\frac{d\mathbf{v}}{dt} = q(\mathbf{E} + \mathbf{v} \times \mathbf{B})$$

$$\therefore \qquad v_x\left[1 - \frac{\omega_c^{\,2}}{\omega^2}\right] = \frac{iqE_x}{m\omega} \qquad \text{where} \qquad \omega_c = \frac{Bq}{m}$$

so $$v_{x1}\left[1 - \frac{\omega_{c1}^{\,2}}{\omega^2}\right] = \frac{q_1}{\omega^2\epsilon_o m_1}(n_1 q_1 v_{x1} + n_2 q_2 v_{x2}) \qquad (13.15)$$

and $$v_{x2}\left[1 - \frac{\omega_{c2}^{\,2}}{\omega^2}\right] = \frac{q_2}{\omega^2\epsilon_o m_2}(n_1 q_1 v_{x1} + n_2 q_2 v_{x2}) \qquad (13.16)$$

Let $\omega_p^{\,2} = \dfrac{nq^2}{\epsilon_o m}$ where ω_p is the ion plasma frequency

Combining (13.15) and (13.16), we obtain

$$\omega^4 - \omega^2(\omega_{c1}^2 + \omega_{c2}^2 + \omega_{p1}^2 + \omega_{p2}^2) + \omega_{c1}^2\omega_{c2}^2 + \omega_{p1}^2\omega_{c2}^2 + \omega_{p2}^2\,\omega_{c1}^2 = 0 \qquad (13.17)$$

For $n \geq 10^{19}$ m^{-3} and $B \sim 1$ Tesla, $\omega_p \gg \omega_c$. There are two solutions of Eq.(13.17). The lower ω solution can be found by ignoring ω^4 in which case

$$\omega^2 = \frac{\omega_{c1}^2\omega_{p2}^2 + \omega_{c2}^2\omega_{p1}^2}{\omega_{p1}^2 + \omega_{p2}^2}$$

which agrees with Eq.(13.12) for a H–D mixture. Using the same approximations,

$$\frac{v_{x1}}{v_{x2}} = -\frac{n_2 q_2}{n_1 q_1} \qquad v_{y1} = \left[\frac{\omega_{c1}}{i\omega}\right] v_{x1} \qquad v_{y2} = \left[\frac{\omega_{c2}}{i\omega}\right] v_{x2}$$

Hence E_x is very small (zero in this approximation) and both the displacement and conduction currents are small. Since v_x and v_y are 90° out of phase and nearly equal in amplitude, the resultant fluid motion is as shown in Fig. 13.6. This type of motion can be set up by displacing a fluid element or by sending a wave through the plasma. Otherwise, the ions rotate in cyclotron orbits without any net fluid motion.

The other solution of (13.17), when $\omega_p \gg \omega_c$, is

$$\omega^2 = \omega_{p1}^2 + \omega_{p2}^2$$

This solution is spurious since it does not correspond to any known wave resonance. It is spurious due to the neglect of any change in the electron number density. However, at very low densities, where $\omega_p \sim \omega_c$ the second solution of (13.17) corresponds to the lower hybrid resonance in a two ion species plasma.

13.7 Minority deuterium in hydrogen

In the minority heating scheme, the minority species is heated at its fundamental cyclotron frequency. The case of minority hydrogen in deuterium was described above. Now suppose that deuterium is the minority species and hydrogen is the majority. If the wave frequency is chosen to coincide with the deuterium cyclotron frequency, then one can expect significant heating from two effects. Firstly, deuterium ions should be heated by cyclotron damping. Secondly, the wave frequency is well below the hydrogen cyclotron frequency and the torsional wave can propagate over a wide cross-section of the plasma. Within this region, wave heating would be expected at an Alfven resonance surface.

The $k_\perp^2 = \infty$ resonance surfaces in a toroidal plasma with $n_d/n_h \sim 0.05$ are shown in Fig. 13.7. These surfaces were calculated for a parabolic density profile using Eq.(13.6) and using local values of $k_\parallel = n/R$ and $B = B_oR_o/R$, where R_o is the major radius. The $\omega = \omega_{cd}$ layer was assumed to be located at $R = R_o$. For typical tokamak parameters, it is necessary to chose a relatively large value of n, typically $n > 5$, to place the resonance surfaces away from the edge of the plasma. This is because $\lambda_\parallel \sim 2\pi v_A/\omega$ is small at high ω. In the conventional Alfven wave heating scheme, ω is well below the cyclotron frequency and $\lambda_\parallel \sim 2\pi R_o$.

As shown in Fig. 13.7, an Alfven resonance layer exists on the high field side of the torus, $R < R_o$, intercepting the $\omega = \omega_{cd}$ layer at the plasma edge. The ion – ion hybrid resonance layer is close to the $\omega = \omega_{cd}$ layer near the centre of the plasma but it evolves into an Alfven resonance layer at large R. The latter effect can be seen more clearly by taking the limit $n_d/n_h \to 0$. In this limit, the two $k_\perp^2 = \infty$ surfaces shown in Fig. 13.7 join smoothly into a single Alfven resonance surface near the plasma edge.

A heating scheme involving deuterium minority in hydrogen is not expected to be successful, since most of the wave energy will be dissipated in Alfven resonance layers near the plasma edge. The only way to avoid edge heating would be to chose an impractically high value of n, say $n = 10$, avoiding excitation of all modes with $n < 10$.

An important question, still to be resolved, is the precise shape of the resonance surfaces in toroidal geometry. Since Alfven and ion–ion hybrid waves are magnetically guided along field lines, the resonance surfaces should coincide with flux surfaces and not cross flux surfaces as shown in Fig. 13.7. This problem is discussed by HELLSTEN and TENNFORS, I3. It is possible, however, that mode conversion might occur along the surfaces shown in Fig. 13.7, but the mode-converted wave will then carry energy away from the conversion region and along flux surfaces. Heating would then be expected over a large volume of the plasma, rather than at localised surfaces.

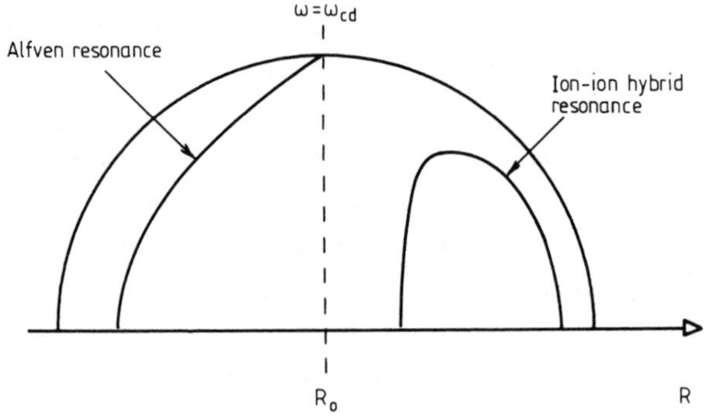

Fig. 13.7 $k_\perp^2 = \infty$ resonance surfaces in toroidal geometry in a hydrogen-deuterium plasma, with $n_d/n_h \sim 0.05$ and with the $\omega = \omega_{cd}$ layer passing through the centre of the plasma.

SPACE FILLERS

(a) Show that the sum of a RHCP field $E_R \exp(-i\omega t)$ and a LHCP field $E_L \exp(+i\omega t)$ is elliptically polarised.

(b) Consider steady state excitation of an Alfven wave in a toroidal cavity. Suppose that after one toroidal trip, the wave is attenuated in amplitude by a factor η and delayed by a phase shift ϕ. By summing an infinite number of terms, calculate the steady state amplitude of the cavity mode. The problem is essentially the same as that of calculating the transmission of light through a Fabry-Perot etalon, since a light beam is reflected an infinite number of times within the etalon and a small fraction emerges on each reflection.

14 INHOMOGENEOUS PLASMAS

As described in Chapter 1, the effects of density and magnetic field gradients in a plasma are of paramount importance in all RF heating schemes. Some of these effects have already been described in previous Chapters, using either a density step to model a density gradient, or by using the WKB approximation, as in Section 11.3. In this Chapter, the problem is attacked directly by solving the wave equations for realistic density profiles. The solutions are known as "full wave solutions" in the sense that they are exact rather than approximate solutions of the wave equations. Throughout this Chapter, it is assumed that the steady magnetic field is uniform, mainly to simplify the algebra. It is not difficult to include magnetic field gradients, at least in ideal MHD, although one needs to be careful to check whether the equilibrium magnetic field satisfies the relations $\nabla \cdot \mathbf{B} = 0$ and $\nabla \times \mathbf{B} = \mu_0 \mathbf{J}$. At frequencies near the ion cyclotron frequency, gradients in **B** are more difficult to handle since the ion cyclotron wave propagates on only one side of the cyclotron layer.

The most common density profile encountered in laboratory plasmas is the parabolic profile $\rho = \rho_0(1 - r^2/a^2)$, where a is the plasma radius. Normally, one would chose a cylindrical coordinate system to analyse wave propagation in a cylindrical plasma, but it is sometimes instructive to use a cartesian coordinate system, as shown in Fig. 14.1. In cartesian coordinates, the parabolic profile is given by

$$\rho = \rho_0[1 - (x^2 + y^2)/a^2] \qquad (14.1)$$

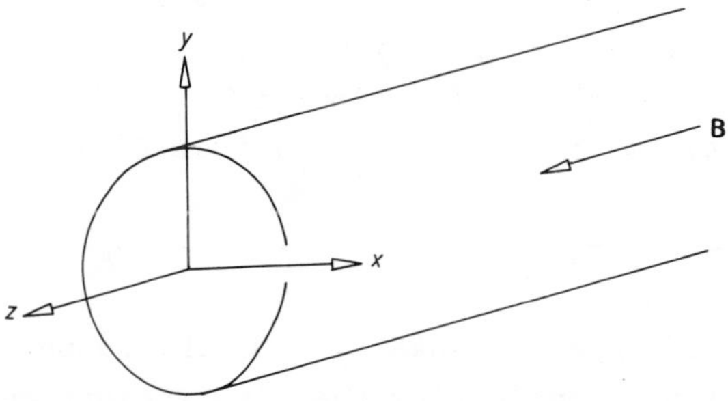

Fig. 14.1 x,y,z coordinate system for a cylindrical plasma.

Another useful profile is the linear density ramp

$$\rho = \rho_0(1 - x/a) \tag{14.2}$$

The ramp profile is a good approximation to a parabolic profile near the plasma edge and has the advantage that ρ varies in the x direction only. A ramp profile in cartesian coordinates is analogous to a parabolic profile in cylindrical coordinates since the density depends on only one coordinate. To simplify the analysis in this Chapter, we make the following assumptions:

1. A uniform field $\mathbf{B} = \hat{\mathbf{z}}\, B$ exists in the z direction.
2. The plasma is uniform in the z direction.
3. The plasma contains only one ion species.
4. There is no steady current in the plasma.
5. The kinetic pressure and resistivity are zero.
6. A conducting wall is located at $r = a$ or at $x = a$.

To simplify the problem even further, we will first consider the ideal MHD equations. Effects of finite frequency will be examined in Section 14.7.

14.1 Ideal MHD equations

Low frequency Alfven waves in an inhomogeneous plasma can be described by the ideal MHD equations,

$$\mathbf{E} + \mathbf{v} \times \mathbf{B} = 0 \qquad \rho\partial\mathbf{v}/\partial t = \mathbf{j} \times \mathbf{B} \tag{14.3}$$

where $\rho(x,y)$ is the equilibrium mass density, $\mathbf{E}$ is the wave electric field, $\mathbf{v}$ is the plasma velocity and $\mathbf{j}$ is the current density. The equilibrium values of $\mathbf{E}$, $\mathbf{v}$ and $\mathbf{j}$ are all assumed to be zero. For sinusoidal perturbations of the form $\exp(-i\omega t)$, we find from (14.3) that $E_z = 0$, $v_z = 0$ and

$$\mu_0 j_x = -i\omega E_x/v_A^2 \qquad \mu_0 j_y = -i\omega E_y/v_A^2 \tag{14.4}$$

where $v_A^2 = B^2/(\mu_0\rho)$. Since $E_z = 0$, the components of Maxwell's equations, neglecting the displacement current, are

$$i\omega b_x = -\frac{\partial E_y}{\partial z} \qquad \mu_0 j_x = \frac{\partial b_z}{\partial y} - \frac{\partial b_y}{\partial z} \qquad (14.5)$$

$$i\omega b_y = \frac{\partial E_x}{\partial z} \qquad \mu_0 j_y = \frac{\partial b_x}{\partial z} - \frac{\partial b_z}{\partial x} \qquad (14.6)$$

$$i\omega b_z = \frac{\partial E_y}{\partial x} - \frac{\partial E_x}{\partial y} \qquad \mu_0 j_z = \frac{\partial b_y}{\partial x} - \frac{\partial b_x}{\partial y} \qquad (14.7)$$

and $$\nabla\cdot\mathbf{b} = \partial b_x/\partial x + \partial b_y/\partial y + \partial b_z/\partial z = 0 \qquad (14.8)$$

From equations (14.4) to (14.6) we find, on differentiating with respect to z, that

$$\frac{\partial^2 b_x}{\partial z^2} + \frac{\omega^2}{v_A{}^2} b_x = \frac{\partial}{\partial z}\left[\frac{\partial b_z}{\partial x}\right] \qquad (14.9)$$

$$\frac{\partial^2 b_y}{\partial z^2} + \frac{\omega^2}{v_A{}^2} b_y = \frac{\partial}{\partial z}\left[\frac{\partial b_z}{\partial y}\right] \qquad (14.10)$$

Equations (14.8) to (14.10) describe arbitrary, small amplitude waves of angular frequency ω. Solutions of this set depend on whether b_z is finite or zero. We first consider solutions with $b_z = 0$. Solutions with finite b_z are considered in Sections 14.3 to 14.7.

14.2 Torsional wave solutions

If $b_z = 0$ for all x,y then (14.9) and (14.10) reduce to

$$\frac{\partial^2 b_x}{\partial z^2} + \frac{\omega^2}{v_A{}^2} b_x = 0 \qquad \frac{\partial^2 b_y}{\partial z^2} + \frac{\omega^2}{v_A{}^2} b_y = 0$$

These equations can be solved by Fourier analysis in the z direction, replacing $\partial/\partial z$ by ik_z, to give

$$\omega/k_z = v_A(x,y) \qquad (14.11)$$

This relation is easily recognised as the dispersion relation for the torsional Alfven wave. It indicates that the phase velocity is equal to the local Alfven speed at all points in the plasma. Despite the fact that (14.11) is the simplest dispersion relation one could possibly imagine in an inhomogeneous plasma, its simplicity is deceptive. In Fourier theory, k_z is a constant wavenumber, independent of x and y. The left side of (14.11) is therefore a constant, but the right side is a function of x and y. There are two distinct interpretations of this result, depending on experimental conditions.

If ω is fixed by an external generator and if k_z is fixed by periodic boundary conditions (eg by toroidal geometry or by a periodic array of antennas), then the torsional wave can exist only at an isolated surface in the plasma where ω/k_z is equal to the local Alfven speed. For a parabolic profile, the contours of constant v_A are circles around the $r = 0$ origin. The wave structure is therefore a δ-function in the r direction. Since $\nabla \cdot \mathbf{b} = 0$, b_r must be zero, so the only wave magnetic field component is b_θ, which is constant in the θ direction since $\partial b_\theta/\partial\theta = 0$. In other words, a torsional wave with fixed k_z and with $b_z = 0$ can exist only as an $m = 0$ mode at an isolated surface in the plasma.

If, on the other hand, k_z is not fixed and is *allowed* to vary with x and y, then a torsional wave with $b_z = 0$ will contain a continuous spectrum of k_z components. An initially plane wave front will then be distorted into a curved wave front as the wave propagates in the z direction, as observed by CROSS and LEHANE (C2). Such a disturbance is said to be phase – mixed by the density gradient. Since different sections of the wave propagate at different speeds, one might expect the transverse gradients of the wave fields to increase as the wave propagates further away from the antenna, as shown schematically in Fig. 14.2. HEYVAERTS and PRIEST (G7) have suggested that this should lead to strong resistive and viscous damping at large z and may therefore be important as a heating mechanism in the solar corona. However, it is shown in Chapter 15 that an alternative outcome is that the transverse gradients will *decrease* at large z due to strong resistive damping of the high $k_\perp$ components of the disturbance. An important effect of phase – mixing, illustrated in Fig. 14.2, is to enhance the high $k_\perp$ spectral composition of the disturbance as it propagates in the z direction. However, the high $k_\perp$ components are more strongly damped in a resistive plasma. The actual outcome will depend on whether the original disturbance is highly localised (ie contains mainly high $k_\perp$ components) or whether it is more extended, containing only low $k_\perp$ components.

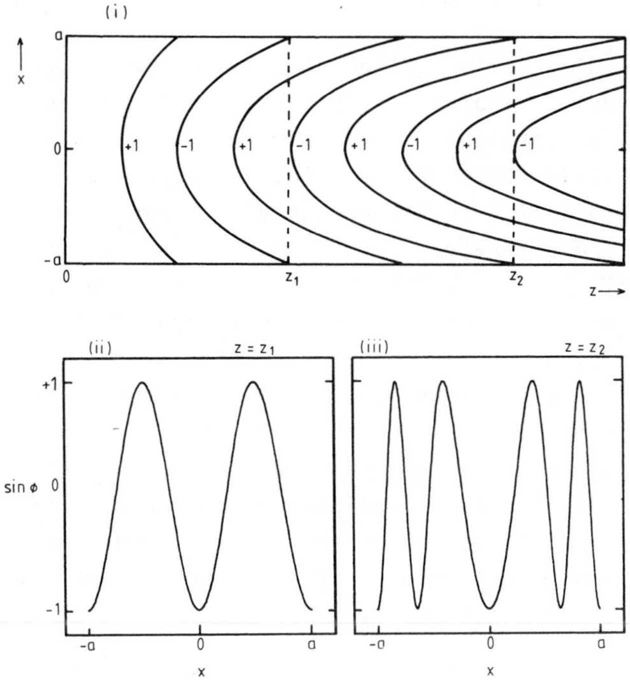

Fig. 14.2 (i) Schematic diagram showing wave fronts (constant phase surfaces, $\sin\phi = \pm 1$) in the x–z plane when k_z varies with x. (ii) $\sin\phi$ vs x at $z = z_1$ (iii) $\sin\phi$ vs x at $z = z_2$.

One approach to analyse a phase-mixed disturbance would be to integrate over the spectrum of k_z components generated by a given antenna. A much simpler method is to assume solutions of the form

$$b \sim f(x,y)\, e^{i[k_z(x,y)z - \omega t]} \tag{14.12}$$

Then $\partial/\partial z = ik_z(x,y)$ and the solution of (14.9) or (14.10), when $b_z = 0$, is

$$\boxed{\omega/k_z(x,y) = v_A(x,y)} \qquad (14.13)$$

The method used to derive (14.13) is not strictly Fourier analysis, since k_z is assumed to vary with x and y in (14.12). The subtle difference between the two methods will be more evident when we consider solutions with finite b_z. Solutions of the form (14.12) are found only when $b_z = 0$.

Several experimental conditions can be recognised where k_z can vary with x and y. For example, in astrophysical plasmas, a torsional wave may propagate along steady magnetic field lines of essentially infinite length. In low temperature cylindrical laboratory plasmas, the torsional wave will be strongly damped, and will not reflect off the far end of the device. In this case there is no k_z selection since wave damping prevents the formation of standing waves in the z direction. Another situation is the case shown in Fig. 6.6 where the torsional wave was excited by a small dipole loop antenna at the edge of a tokamak plasma. The wave propagated as a narrow beam along helical magnetic field lines in the edge plasma and did not return to the antenna since it was located on an irrational magnetic flux surface. Again, no standing wave was formed and the parallel wavelength was not fixed by any periodic boundary condition.

If k_z is allowed to vary with x and y, and if we assume perturbations of the form given by (14.12), then equations (14.5) to (14.8), with $b_z = 0$, give

$$\frac{\partial}{\partial x}\left[\frac{b_x}{k_z}\right] + \frac{\partial}{\partial y}\left[\frac{b_y}{k_z}\right] = 0$$

$$\therefore \qquad \boxed{\mathbf{b} \cdot \nabla k_z = b_x \frac{\partial k_z}{\partial x} + b_y \frac{\partial k_z}{\partial y} = 0} \qquad (14.14)$$

Since the gradient of k_z is perpendicular to **b**, the wave magnetic field lines form contours of constant k_z. This places a restriction on the allowed mode structure in the x–y plane. The ratio of the magnetic field components is given, from (14.13) and (14.14), by

$$\frac{b_y}{b_x} = -\frac{\partial k_z/\partial x}{\partial k_z/\partial y} = -\frac{\partial v_A/\partial x}{\partial v_A/\partial y}$$

Consider, for example, the ramp profile (14.2) where v_A is independent of y. Then $b_x = 0$ and hence $E_y = v_x = 0$. In this case, each layer of plasma shears past adjacent layers without interaction. From (14.8), we see that $\partial b_y/\partial y = 0$, corresponding to a $k_y = 0$ mode. For the parabolic density profile $\rho = \rho_0(1 - r^2/a^2)$, where $r^2 = x^2 + y^2$,

$$\frac{b_y}{b_x} = -\frac{\partial \rho/\partial x}{\partial \rho/\partial y} = -\frac{x}{y}$$

The wave magnetic field lines are therefore concentric circles around the $r = 0$ origin. The same magnetic field structure is found for any other azimuthally symmetric density profile. In cylindrical coordinates, the only nonzero magnetic field component is b_θ, and the only velocity component is v_θ. Since $\nabla\cdot\mathbf{b} = 0$, $\partial b_\theta/\partial\theta = 0$, corresponding to an $m = 0$ mode. Each concentric annulus undergoes purely torsional oscillations, independently of oscillations in adjacent annuli. The radial structure of b_θ is arbitrary, but will of course depend on the structure of the launching antenna. For example, a narrow beam will propagate along the $r = 0$ axis if it is generated by a small antenna located at $r = 0$, as shown in Fig. 6.1. These results were first derived by PNEUMAN (B10) by considering the azimuthally symmetric $m = 0$ mode in a cylindrical plasma. However, he did not show that the $m = 0$ mode is the *only* mode of oscillation of the torsional wave when $b_z = 0$. In Section 14.6 we will consider torsional oscillations about a displaced origin, in which case $b_z \neq 0$. It will then become apparent that the $m = 0$ mode is the *only* mode which can be described by equations (14.13) or (14.14). The latter result has a simple physical interpretation. Adjacent, concentric annuli behave independently only when undergoing torsional oscillations about the centre of symmetry of the equilibrium density distribution. Any $m \neq 0$ disturbance will involve fluid motion across adjacent annuli and hence the annuli will be coupled.

14.3 Compressional wave solutions

When b_z is finite, there are no solutions of (14.9) or (14.10) of the form given by equation (14.12). To prove this assertion, suppose that b_z and b_x do vary according to (14.12). Then

$$\frac{\partial b_z}{\partial x} = \left[\frac{\partial f}{\partial x} + izf\,\frac{\partial k_z}{\partial x}\right] e^{i(k_z z - \omega t)}$$

so equation (14.9) has the form

$$b_x = \frac{i}{F}\left[k_z\,\frac{\partial f}{\partial x} + (1+ik_z z)f\,\frac{\partial k_z}{\partial x}\right] e^{i(k_z z - \omega t)} \qquad (14.15)$$

where $F = \omega^2/v_A{}^2 - k_z{}^2$. The amplitude term in (14.15) is now a function of z, which is inconsistent with the original assumption that b_x should vary according to (14.12).

We conclude that k_z can vary with x and y only when $b_z = 0$. This result is consistent with Snell's law, equation (7.2). Snell's law indicates that k_z is continuous across the boundary separating two uniform density regions with a density jump in the x direction, unless b_x and b_z are both zero. These are the components required to satisfy the boundary conditions. Both components are zero for the torsional wave when $k_y = 0$. Similarly, both b_r and b_z are zero for the m = 0 mode in a cylindrical plasma. Consequently, it is only the $k_y = 0$ torsional mode in a plasma slab, or the m = 0 torsional mode in a cylinder, which is phase-mixed by a density gradient.

Since k_z is independent of x, y and z when b_z is finite, we can Fourier analyse (14.8) to (14.10) in the z direction, replacing $\partial/\partial z$ by ik_z to give

$$b_x = \frac{ik_z}{F}\,\frac{\partial b_z}{\partial x} \qquad\qquad b_y = \frac{ik_z}{F}\,\frac{\partial b_z}{\partial y} \qquad (14.16)$$

where

$$F = \frac{\omega^2}{v_A{}^2} - k_z{}^2 \qquad (14.17)$$

Substituting (14.16) in (14.8), we obtain

$$\frac{\partial}{\partial x}\left[\frac{1}{F}\frac{\partial b_z}{\partial x}\right] + \frac{\partial}{\partial y}\left[\frac{1}{F}\frac{\partial b_z}{\partial y}\right] + b_z = 0 \qquad (14.18)$$

The simplest solution of this equation is obtained when the plasma is homogeneous so $\partial F/\partial x = 0$ and $\partial F/\partial y = 0$. If we replace $\partial/\partial x$ by ik_x and $\partial/\partial y$ by ik_y then

$$F = k_x^2 + k_y^2 = \omega^2/v_A^2 - k_z^2$$

This is the dispersion relation for the compressional wave. Now consider the case where there is a density gradient in the x direction only and where $\partial/\partial y = ik_y$ (the equivalent problem in cylindrical geometry is one where the density gradient is in the r direction and $\partial/\partial\theta = im$). Then (14.18) takes the form

$$\frac{\partial^2 b_z}{\partial x^2} + \left[\frac{2\omega^2}{Fv_A^3}\frac{\partial v_A}{\partial x}\right]\frac{\partial b_z}{\partial x} + (F - k_y^2)b_z = 0 \qquad (14.19)$$

This equation can be solved only by numerical methods. There is no problem in obtaining such a solution when $k_y = 0$ and an example is given in Fig. 14.3. However, when $k_y \neq 0$, Eq.(14.19) is singular at the location $x = x_o$ where $F = 0$. This is the location of the Alfven resonance layer. Analytic solutions can be obtained in the neighbourhood of the singularity and matched to the numerical solutions either side of the resonance layer. The latter method of solution is considered in Section 14.4.

Equation (14.19) is of the general form

$$d^2y/dx^2 + 2\gamma dy/dx + k_x^2 y = 0$$

and therefore describes damped compressional wave oscillations with a damping term, γ, proportional to $\partial v_A/\partial x$. If $\partial v_A/\partial x$ is small then the solution of (14.19) is approximately

$$b_z = b_o e^{-\gamma x} e^{ik_x x} \qquad (14.20)$$

where $\gamma = \dfrac{\omega^2}{Fv_A{}^3}\dfrac{\partial v_A}{\partial x}$ and $k_x{}^2 = F - k_y{}^2$

This solution is only approximate since both γ and k_x vary with x. It is a good approximation if the density varies slowly on a scale compared with $\lambda_x = 2\pi/k_x$. This is the WKB approximation. It allows one to set $\partial(ik_x x)/\partial x = ik_x$ by assuming that the term $x\ \partial k_x/\partial x$ is negligible.

The "damping" term introduced by the density gradient does not correspond to dissipation of wave energy. The variation of the wave amplitude in the x direction is actually due to refraction of the compressional wave. Despite the fact that (14.20) appears to indicate wave damping, note that if $\partial v_A/\partial x$ is negative, then (14.20) describes a growing wave. Since k_z is assumed to be constant and since k_x varies with x, the propagation vector changes direction with x. The direction of refraction depends on whether the wave propagates towards or away from the high density region.

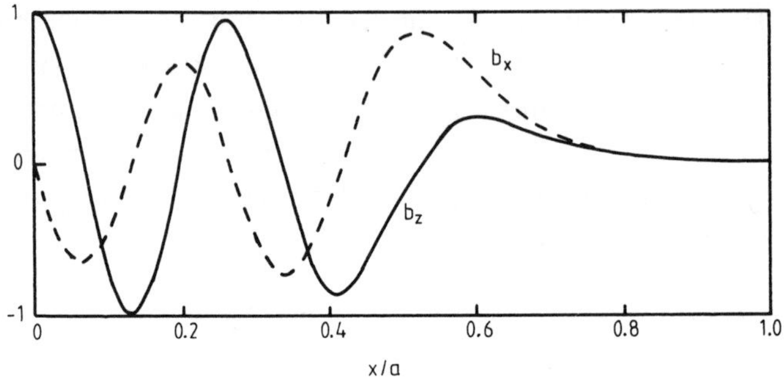

Fig. 14.3 Numerical solution of Eq. (14.19) for a hydrogen plasma with B = 1.0 T, $n = n_0(1 - x^2/a^2)$, $n_0 = 1 \times 10^{21} m^{-3}$, a = 0.5 m, f = 9.2 MHz, $k_y = 0$ and F = 0 at x/a = 0.6.

The solution (14.20) is clearly wrong when γ is very large. In fact, for the conditions shown in Fig. 14.3, F = 0 at x/a = 0.6 and then $\gamma = \infty$. When F = 0, both the torsional and compressional wave types have the same ω and k_z.

Normally, this leads to strong coupling between the two wave types. However there is no coupling when $k_y = 0$ since then $b_y = 0$ for the compressional wave and $b_x = b_z = 0$ for the torsional wave. In order for coupling or mode conversion to occur, both wave types need at least one common wave field component, as described in Chapter 7.

The compressional wave shown in Fig. 14.3 is well behaved at F = 0, but not all wave frequencies will satisfy the boundary conditions. At 9.2 MHz the solution satisfies the boundary condition $b_x(a) = 0$ if a conducting wall is located at $x = a$. For a given value of k_z, there is a discrete set of frequencies where boundary conditions are satisfied, corresponding to different numbers of nodes between $x = 0$ and $x = a$. An inhomogeneous plasma therefore possesses a discrete spectrum of compressional wave modes.

An inhomogeneous plasma also possesses a continuous spectrum of torsional wave modes, at least in the ideal MHD limit. By this we mean that if k_z is fixed (for example by an external set of antennas) then there will be a continuous range of frequencies where the boundary conditions are satisfied. The continuous spectrum is described in the following Section.

14.4 The Alfven resonance

When $k_y \neq 0$, it is not possible to obtain a solution of the type shown in Fig. 14.3 since b_x and $b_y \rightarrow \infty$ as $F \rightarrow 0$. This is due to mode conversion of the compressional wave to a torsional wave at the Alfven resonance surface where $\omega = k_z v_A(x)$. However, it is still possible to construct a solution which satisfies boundary conditions by joining numerical solutions of Eq. (14.19) away from the resonant surface to analytical solutions near the surface. The boundary condition $b_x = 0$ at a conducting wall is then satisfied, but $b_x \rightarrow \infty$ from both sides of the resonant surface as shown schematically in Fig. 14.4. Even though the equations are no longer linear when $b_x \rightarrow \infty$, and realistic effects such as finite resistivity will prevent b_x increasing to infinity, we will explore these solutions for their intrinsic interest. At least they indicate that b_x will become large under realistic plasma conditions.

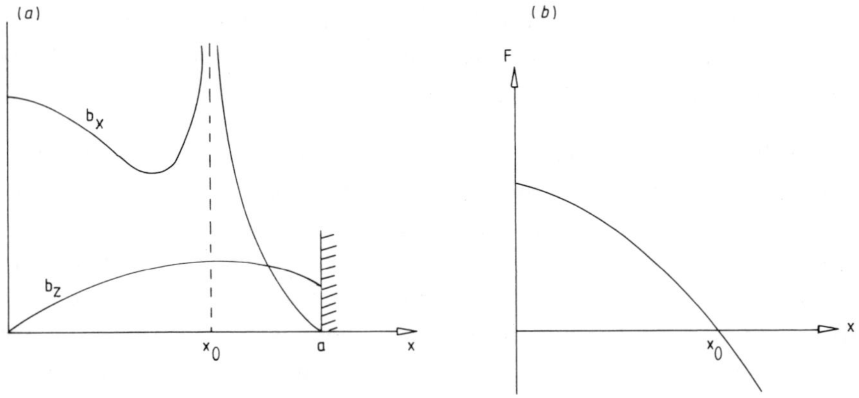

Fig. 14.4 (a) Solution of Eq. (14.19) at low wave frequency with $k_y \neq 0$ and with an Alfven resonance layer at $x = x_0$. (b) F vs x.

Suppose that $F = 0$ at $x = x_0$. Then for x close to x_0, let $F = -Av$ where $v = x - x_0$ and $A = 2\omega^2(\partial v_A/\partial x)/v_A{}^3$. Equation (14.19) can then be expressed as

$$\frac{\partial^2 b_z}{\partial v^2} - \frac{1}{v}\frac{\partial b_z}{\partial v} - (Av + k_y{}^2)b_z = 0 \tag{14.21}$$

Equation (14.21) has a regular singularity at $v = 0$ with two independent series expansion solutions

$$b_z = b_0(v^2 + k_y{}^2\frac{v^4}{8} + \frac{A}{15}v^5 + \dots) \tag{14.22}$$

or $$b_z = b_1(1 + k_y{}^2\frac{v^2}{2}\ell n\, v + \dots) \tag{14.23}$$

Substitution into (14.21) will verify that these are good approximations when v is small. The question arises as to the nature of the second solution when $v < 0$. An obvious approach is to change the variable v in (14.21) to $u = -v$, in which case v in (14.23) is replaced by u. However, this approach does not connect the solutions either side of $v = 0$. Analytic continuation of the solutions across $v = 0$ requires that for $v < 0$,

$$\ell n\ v = \ell n\ |v| + i\pi$$

since $e^{i\pi} = -1$. The imaginary contribution to b_z means that the wave is damped, even though there is no explicit damping mechanism included in the ideal MHD equations. The rate of damping can be determined by integrating the Poynting flux in the x direction (E5, M17). This particular aspect of the problem has led to a great deal of discussion in the past, both in relation to the Alfven heating scheme and concerning the analogous problem of resonant absorption of an electromagnetic wave in an inhomogeneous plasma at the local electron plasma frequency. Nevertheless, the solution is plausible from the point of view that some form of dissipation is required to remove the unphysical singularity in the equations and this will result in wave damping. Indeed, one finds that inclusion of realistic damping results in a wave damping rate which is independent of the amount of damping assumed (D6,E5,H7,H15), provided the amount of damping is small. For example, if finite resistivity is included as a dissipation mechanism, then as the resistivity is reduced, the resonance amplitude is increased but the width of the resonance layer is reduced, with the result that the total energy absorption rate is unchanged.

It is quite amazing that the magnitude of the wave damping can be determined even when there is no damping mechanism in the equations. There is an analogy, of sorts, in electrical circuit theory. If a charged capacitor is discharged into another, uncharged capacitor, the total charge will be conserved but the total stored energy is reduced. The missing energy can be accounted for by allowing the connecting leads to have resistance R. It is then easy to show, from the time integral of Ri^2 that the total energy dissipated in R is independent of R, even when $R \to 0$.

A particular solution of (14.19) satisfying boundary conditions will require a linear combination of the independent solutions (14.22) and (14.23) either side of the resonance layer. This method of solution is outlined by McPHERSON and PRIDMORE-BROWN (C11) and numerical results are given by KIVELSON and SOUTHWOOD (M7). As $v \to 0$, b_z remains finite but $b_x \propto \ell n(x-x_0)$ and $b_y \to 1/(x-x_0)$ from (14.16) and (14.23). At $x = x_0$, the wave magnetic field is therefore polarised in the resonance surface.

McPHERSON and PRIDMORE-BROWN were among the first of many authors to note that the boundary conditions can be satisfied, at fixed k_z, over a continuous range of frequencies. This frequency range is commonly known as the Alfven continuum. Within this range, the Alfven resonance surface moves continuously, as the frequency is increased, from the centre of the plasma (small v_A) to the edge (large v_A). We note in passing that, in a current carrying plasma, the resonance surface is defined by $\omega = k_\parallel v_A$ rather than by $\omega = k_z v_A$. Since $k_\parallel$ is also a function of r in general, the Alfven resonance surface, at the low frequency end of the continuum, is not necessarily located at the centre of the plasma. The location of the resonance layer, in a current carrying plasma, is given by equation (11.27).

A feature of the Alfven resonance is that the wave magnetic field component parallel to the resonance surface, b_y, reverses sign across the surface. This result follows from (14.16), giving $b_y = -k_y k_z b_z/F$. Since b_z is continuous across the surface but F reverses sign, then b_y reverses sign. As shown in Fig. 14.4, b_x does not reverse sign. Therefore, the polarisation (ie the sign of ib_x/b_y) reverses across the resonance surface. The sense of rotation of the wave fields depends on the sign of k_y or, in cylindrical geometry, on the sign of m. As shown in Fig. 9.2, one cannot assume that the torsional wave is always left-hand polarised. The sense of rotation of the wave fields provides an important piece of experimental evidence in identifying an Alfven resonance condition, as described in Section 14.5.

Equivalent solutions can be obtained from the equation for b_x. With $\partial/\partial y = ik_y$, we find from (14.16) and the x derivative of (14.18) that

$$\frac{\partial}{\partial x}\left[\frac{F}{F - k_y{}^2}\,\frac{\partial b_x}{\partial x}\right] + Fb_x = 0$$

Near F = 0,
$$\frac{\partial^2 b_x}{\partial v^2} + \frac{1}{v}\,\frac{\partial b_x}{\partial v} - k_y{}^2\, b_x = 0$$

This is a standard Bessel equation whose solution is

$$b_x \;=\; b_0 I_0(k_y v) + b_1 K_0(k_y v)$$

The I_0 and K_0 functions can be represented in series form by

$$I_0(x) = 1 + \frac{x^2}{4} + \frac{x^4}{64} + \;\ldots \qquad (x \ll 1)$$

$$K_0(x) \;=\; \frac{x^2}{4} - \left[0.577 + \ell n_e\left(\frac{x}{2}\right)\right] I_0(x) + \ldots \;(x \ll 1)$$

so $K_0(x) \to \infty$ as $x \to 0$. The $K_0(x)$ function is analytically continued across x = 0 by the relation

$$K_0(-x) \;=\; K_0(x) + i\pi\, I_0(x)$$

Solutions of this form are described by STIX (A8, Ch 10) in the more general context of wave propagation in plasmas with a zero or infinity in the refractive index.

A relatively simple case of Alfven resonance damping arises for surface waves in a plasma with a finite width surface. If the surface is infinitely thin, as shown in Fig. 14.5a, and if the plasma density is uniform in the half space $x < 0$, then the plasma will support a surface wave which is undamped and propagates at a phase velocity $\omega/k_z = \sqrt{2}v_A$. Now consider a finite width surface with a linear density ramp as shown in Fig. 14.5b. If the surface width, a, is small, then the phase velocity of the surface wave is not significantly different from $\sqrt{2}v_A$. The local Alfven speed in the ramp varies from v_A at $x = -a$ to ∞ at $x = 0$. Hence there will be a surface within the ramp where the phase velocity of the surface wave is equal to the local Alfven speed. As a result, the surface wave is damped and the plasma will be locally heated. This problem is treated analytically by several authors (D6, D10, E5) who show that the damping rate approaches zero as $a \to 0$. For any given a, the damping rate is independent of the plasma resistivity, as described above.

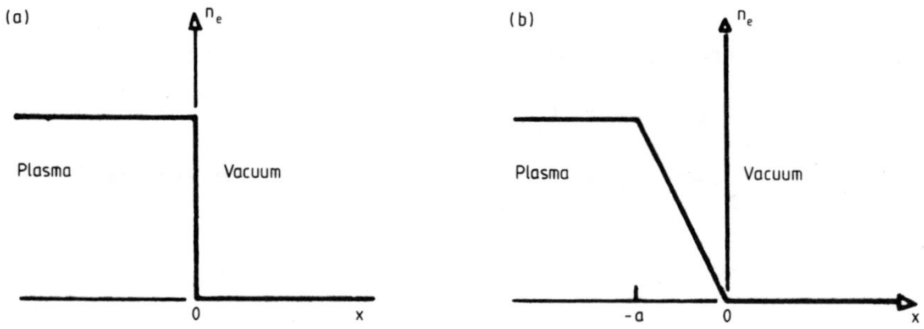

Fig. 14.5 Density profiles which support (a) undamped and (b) damped surface waves.

14.5 Geomagnetic micropulsations

Theoretical and experimental work on the Alfven resonance has developed along parallel lines, largely independently, in three separate areas. Work in laboratory plasmas and the problem of coronal heating in the Sun were mentioned briefly in Chapter 1. A third major area of research is concerned with geomagnetic micropulsations (JACOBS, M6). These are small amplitude (~1nT), very low frequency (~10 mHz) fluctuations in the Earth's magnetic field, generated by the solar wind or by wave–particle interactions in the magnetosphere (M19).

The solar wind generates Alfven surface waves propagating along the magnetopause boundary, (Fig. 1.3), due to the Kelvin–Helmholtz instability. The same effect is responsible for wind–driven waves at a water surface. The wave propagates in the same direction as the wind. Mode conversion of the surface wave can occur at an Alfven resonance surface located between the magnetopause and the Earth. In this geometry, the resonance surface is a magnetic shell formed by the Earth's magnetic field lines. The torsional wave generated at the resonance surface is guided along the Earth's magnetic field lines and is reflected at the Earth's surface to form a standing wave between conjugate points in the northern and southern hemispheres.

There is a considerable amount of experimental evidence in support of Alfven resonance excitation of torsional waves in the magnetosphere. The most convincing evidence is found from observations of the wave field polarisation either side of the resonance layer (M17, M18). Regions either side of the resonance layer, where the wave field is a maximum, are observed to be oppositely polarised. Furthermore, the sense of polarisation reverses at about midday, corresponding to a reversal in the sign of the transverse wavenumber, k_y. The results are consistent with waves propagating westwards in the morning and eastwards in the afternoon, as expected if surface waves are generated by the solar wind.

The presence of standing Alfven waves along the Earth's magnetic field lines is confirmed by simultaneous measurements at each end of the field line and with satellites located on the same field line. These measurements can be made either with a flux-gate or Zeeman effect magnetometer or by a search coil of say 100,000 turns wound on a mumetal core. TAKAHASHI and McPHERRON (M19) summarise data observed for the fundamental and harmonic modes, up to the sixth harmonic. These field line resonances are closely analogous to the vibrations of stretched strings and provide valuable ground-based information on conditions in the magnetosphere.

Not all field line resonances can be explained in terms of mode conversion of a surface wave. This explanation is satisfactory only for low values of k_y. It is frequently observed that field line resonances are highly localised in both directions transverse to the steady magnetic field. A localised mode of this type contains a continuous spectrum of high k_y components. Under these conditions, the surface wave is highly evanescent in a direction parallel to the steady magnetic field and is therefore decoupled from the torsional wave. Propagation of a strongly localised torsional wave is described in the following Section.

14.6 Guided wave solutions

In addition to the singular solutions of Eq. (14.18), described in Section 14.4, there also exist *non-singular* solutions of equation (14.18). Non-singular solutions arise if $\partial b_z/\partial x$ and $\partial b_z/\partial y$ are both zero when F = 0. The nonsingular solutions describe magnetically guided torsional waves which are decoupled from the compressional wave since there is no mode conversion or resonance surface.

An important difference between Alfven singular (continuum) modes and guided Alfven modes is the direction of the wave magnetic field at the F = 0 surface. In a continuum mode, **b** lies in the surface. The direction of **b** in a guided mode can be found as follows. From equation (14.16), $(\mathbf{b} \times \nabla b_z)_z = 0$ at all points in the plasma cross-section. The wave magnetic field lines in the x–y plane are therefore perpendicular to the constant b_z contours. Since $\partial b_z/\partial x$ and $\partial b_z/\partial y$ are both zero at F = 0 for a nonsingular mode, the F = 0 surface forms a constant b_z contour. The wave magnetic field lines therefore cross the F = 0 surface at right angles.

Consider, for example, the wave magnetic field distribution shown in Fig. 14.6. As shown in the figure, the wave magnetic field lines in the x – y plane are assumed to be circular about an arbitrary origin at $y = 0$, $x = x_0$ where the density gradient is finite. The origin could therefore be located near the edge of a cylindrical plasma, but not at the centre, otherwise the mode would be the m = 0 mode studied in Section 14.2. This field distribution is considered because it represents, in the case of a homogeneous plasma, the torsional wave field distribution generated by a current element parallel to the steady magnetic field (BORG et. al., G1). As shown by these authors, only current elements parallel to **B** can launch a torsional wave at low frequencies. Radial feeder elements (perpendicular to **B**) do not launch the torsional wave unless the wave frequency is a significant fraction of the ion cyclotron frequency.

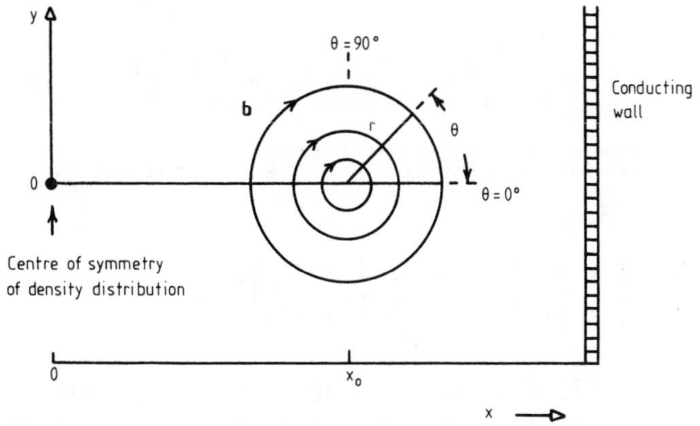

Fig. 14.6 Wave magnetic field lines, **b**, in the x–y plane for a guided torsional wave propagating in the z direction (out of the plane of the page).

An important feature of the field distribution shown in Fig. 14.6 is that it represents a localised disturbance rather than a global or waveguide mode, and therefore contains a continuous spectrum of k_x and k_y components. Alternatively, in cylindrical coordinates, it contains an infinite series of discrete m components when referred to an origin at the centre of symmetry of the density distribution. Another important feature is that the disturbance does not propagate radially outwards from the $y = 0$, $x = x_0$ origin. As shown below, it propagates in the z direction, as a narrow beam or ray, preserving its radial structure.

Since the fluid motion in a torsional wave is parallel to the *wave* magnetic field lines, adjacent annuli centred on an origin at $y = 0$, $x = x_0$ are non-interacting, but the local Alfven speed varies within each annulus. To analyse the wave motion, one could, in principle, apply equation (14.18), and integrate in two dimensions over the plasma cross-section. However, it is more convenient to adopt a cylindrical coordinate system centred at $y = 0$, $x = x_0$, with r and θ defined in Fig. 14.6. Assuming that the wave fields vary as $f(r,\theta)\exp[i(k_z z - \omega t)]$, where k_z is constant, then the equivalent forms of (14.8) and (14.16) are

$$\frac{1}{r}\frac{\partial}{\partial r}(rb_r) + \frac{1}{r}\frac{\partial b_\theta}{\partial \theta} + ik_z b_z = 0 \qquad (14.24)$$

$$b_r = \frac{ik_z}{F}\frac{\partial b_z}{\partial r} \qquad\qquad b_\theta = \frac{ik_z}{rF}\frac{\partial b_z}{\partial \theta} \tag{14.25}$$

where F is defined by (14.17). Substitution of (14.25) in (14.24) gives

$$\frac{1}{r}\frac{\partial}{\partial r}\left(\frac{r}{F}\frac{\partial b_z}{\partial r}\right) + \frac{1}{r^2}\frac{\partial}{\partial \theta}\left(\frac{1}{F}\frac{\partial b_z}{\partial \theta}\right) + b_z = 0 \tag{14.26}$$

Solutions of (14.26) can be sought for any given density profile (to obtain the wave fields) or for any given wave field distribution (to obtain the density profile). It is sufficient, for purposes of illustration, to consider simple cases where (a) $b_r = 0$ or (b) the plasma is inhomogeneous in the x direction only. It turns out that these two cases yield essentially the same results, provided the density varies as a linear ramp in the x direction. We consider case (a), shown in Fig. 14.6, as an example. If $b_r = 0$ for all r,θ then $\partial b_z/\partial r = 0$ from (14.25), and (14.26) reduces to

$$\frac{\partial}{\partial \theta}\left(\frac{1}{F}\frac{\partial b_z}{\partial \theta}\right) + r^2 b_z = 0 \tag{14.27}$$

Equation (14.27) is non–singular at F = 0 and has an exact analytical solution

$$b_z = b_o \sin\theta \tag{14.28}$$

provided that

$$F = -\frac{F_o r\cos\theta}{(1 - F_o r^3\cos\theta)} \tag{14.29}$$

where b_o and F_o are arbitrary constants. From (14.25), the variation of b_θ is given by

$$b_\theta = -\frac{ik_z b_o}{F_o r^2}(1 - F_o r^3\cos\theta) \tag{14.30}$$

Other solutions of (14.27) can be found for different $F(\theta)$ functions or for different $b_z(\theta)$ functions, both of which must be periodic in 2π. The various solutions correspond to different density distributions. Solutions (14.28) and (14.29) are specific for a unique density profile, essentially a linear ramp. Equation (14.29) shows that $F = 0$ at $\theta = \pm 90°$ (ie at $x = x_o$). Since k_z is assumed to be constant, (14.17) is satisfied only if v_A remains constant in the y direction at $x = x_o$. From (14.17) and (14.29), we find that

$$k_z = \omega/v_A(x_o)$$

and

$$\rho = \rho(x_o)\left[1 - \frac{v_A{}^2(x_o)F_o r\cos\theta}{\omega^2(1 - F_o r^3\cos\theta)}\right] \qquad (14.31)$$

where $\rho(x_o)$ and $v_A(x_o)$ are respectively the density and Alfven speed at $x = x_o$. This profile is equivalent to the linear ramp

$$\rho = \rho_o(1 - x/a)$$

if we set $F_o = \omega^2/(v_{Ao}{}^2 a)$, $v_{Ao}{}^2 = B^2/(\mu_o\rho_o)$, $x = x_o + r\cos\theta$ and assume that $F_o r^3 \ll 1$. The assumption $F_o r^3 \ll 1$ is satisfied for most conditions of interest and is always satisfied at sufficiently small r. Consider typical tokamak parameters $v_{Ao} \sim 5 \times 10^6$ m/s, $a \sim 0.3$ m and let $\omega \sim 10^7$ r/s. Then $F_O \sim 13\ m^{-3}$ and $F_O r^3 \ll 1$ when $r < 0.1$ m. For these parameters, $b_\theta \propto 1/r^2$ from (14.30) and $|b_z/b_\theta| \sim F_o r^2/k_z$. For example, at $r \sim 0.02$ m, $|b_z/b_\theta| \sim 0.005$. Homogeneous plasma results are recovered as $a \to \infty$, in which case $F_o \to 0$ and $b_z/b_\theta \to 0$.

Since b_θ is typically much larger than b_z and varies as $1/r^2$, the wave is strongly localised near the $y = 0$, $x = x_o$ origin. If it is localised near the edge of a tokamak plasma, a linear ramp will provide a good approximation to the actual density profile. However, the main point of interest is not the precise shape of the density profile, but the fact that we have now demonstrated that equations (14.18) and (14.26) have non-singular solutions, even when ω lies in the Alfven continuum. The torsional wave described above propagates at the local Alfven speed at $x = x_o$ but not at other values of x.

The above wave field solutions are not physically realistic either at small or large r. As $r \to 0$, $b_\theta \to \infty$ according to (14.30). This behaviour is analogous to the b_θ field surrounding an infinitely thin current carrying wire. The singularity in b_θ at $r = 0$ for the torsional wave can be avoided by considering non ideal MHD effects such as finite resistivity or finite electron mass. This singularity is clearly *not* associated with an Alfven resonance at $x = x_0$.

At large r, the field lines will not remain circular if they intersect a conducting wall boundary. However, this effect is not significant, for two reasons. Firstly, the energy density in the wave fields decreases as $1/r^4$ away from the $x = x_0$ origin, so the wave energy near the boundary is insignificant. Secondly, the wave energy propagates almost exactly along the steady magnetic field lines and not towards the boundary. The direction and magnitude of the energy flux is given by the Poynting vector

$$\mathbf{S} = \frac{\mathbf{E} \times \mathbf{b}}{\mu_0} = \frac{\omega}{\mu_0 k_z}\left[-\hat{\mathbf{x}}(b_x b_z) - \hat{\mathbf{y}}(b_y b_z) + \hat{\mathbf{z}}(b_x{}^2 + b_y{}^2)\right] \qquad (14.32)$$

Equations (14.5) and (14.6) were used to eliminate **E** in (14.32). When $b_z = 0$, **S** is exactly parallel to the steady field. If the wave has a small b_z component, some energy will propagate across the steady field lines. However, as shown above, $b_z / |b_\perp|$ may be well below 0.01 for typical laboratory conditions and the energy flux across field lines is then negligible. Furthermore, the direction of energy flow across the steady field is parallel to $\mathbf{b}_\perp$, where $\mathbf{b}_\perp = \hat{\mathbf{x}}\, b_x + \hat{\mathbf{y}}\, b_y$. The wave magnetic field lines in Fig. 14.6 therefore indicate the direction of energy flow perpendicular to **B**. No energy escapes through surfaces bounded by these curves. It is clear from Fig. 14.6 and the fact that the energy flow is primarily in the z direction that the wave is guided by the steady magnetic field and the plasma boundary has no significant effect on the wave motion. It is also evident that the wave is not refracted by the density gradient.

One could argue that the mode described above should not be described as a torsional wave since (a) it does not locally satisfy the dispersion relation $\omega/k_z = v_A$ at all points in the plasma cross-section and (b) it has a finite b_z component. These properties are more characteristic of the compressional wave. However, the compressional wave does not have its Poynting vector aligned parallel to **B** unless $k_\perp = 0$ (where $k_\perp^2 = k_x^2 + k_y^2$). The mode depicted in Fig. 14.6 has a dominant b_θ component which varies as $1/r^2$ and is therefore a high $k_\perp$ mode, with $k_\perp \gg k_z$. These properties, together with the fact that **S** is essentially parallel to **B**, are all characteristic of the torsional wave.

The fact that a guided beam is decoupled from the compressional wave is not entirely surprising. The dispersion relation for the compressional wave is $k_z^2 = \omega^2/v_A^2 - k_\perp^2$. At high $k_\perp$, the compressional wave is strongly evanescent in the z direction and there is then no coupling between the two wave types. This effect can be regarded as being due to a mismatch of boundary conditions. Both k_z and k_y must be continuous at the boundary separating two uniform density regions with a density jump in the x direction, in order to satisfy boundary conditions at all points in the surface. The boundary conditions do not require k_x to remain continuous. Consequently, mode conversion of a low k_x, low k_y compressional wave to a high k_x torsional wave can occur, provided that both k_y and k_z are continuous at the boundary. However, at large k_y, $k_z^2 < 0$ for the compressional wave and mode conversion cannot occur.

The fact that a guided beam does not locally satisfy the dispersion relation $\omega/k_z = v_A$ at all points is also not entirely surprising. This relation describes the torsional wave in an inhomogeneous plasma only when $b_z = 0$. At finite k_y, b_z is finite and the dispersion relation is slightly different. The condition for an Alfven resonance is also given by the relation $\omega/k_z = v_A$, since at the resonance layer, $b_z/b_x = 0$. At points slightly removed from the resonance layer, b_x varies as $\ln(x - x_0)$ and $\omega/k_z \neq v_A$. Despite the fact that there is a continuous transition from compressional to torsional wave behaviour as $x \to x_0$, one can regard the wave "near" $x = x_0$ as a magnetically guided torsional wave with finite b_z and with $\omega/k_z \neq v_A$. A guided beam has an analogous structure, but b_θ varies as $1/r^2$ rather than logarithmically since the beam contains a spectrum of k_y components.

14.7 Finite frequency effects

To obtain a more realistic model of Alfven waves in laboratory plasmas, we now consider wave propagation at finite frequency in an inhomogeneous plasma. However, for simplicity we assume that (a) there is no steady plasma current (b) there is only one ion species and (c) there is no Alfven resonance layer in the plasma. The resonance layer is absent when $\omega > \omega_{ci}$. It is also absent, when $\omega < \omega_{ci}$, in a narrow frequency range near the lower edge of the continuum. In the absence of a resonance layer, there are no singularities to worry about and the solutions of the wave equations for an inhomogeneous plasma yield a discrete set of waveguide modes. As described in Sect. 11.3, discrete Alfven waves can propagate in a current - carrying plasma. They can also propagate in an inhomogeneous, current - free plasma as a result of finite frequency effects.

A condition for a discrete torsional wave to exist in an inhomogeneous plasma is that the region labelled T in Fig. 11.4 (including the evanescent part) must extend over the whole plasma cross - section so that there is no $k_\perp{}^2 = \infty$ surface anywhere in the plasma. Furthermore, $k_\perp{}^2$ must be sufficiently large that the average value of $k_\perp$, times the plasma radius a, is greater than about 2 or 3. This condition follows by analogy with the homogeneous plasma solutions where the wave fields vary as $J_m(k_\perp r)$. To satisfy boundary conditions at $r = a$, $k_\perp a \sim 2$ or 3 for the first radial waveguide modes. In the ideal MHD limit, discrete torsional wave modes cannot propagate in an inhomogeneous plasma since the region labelled T in Fig. 11.4 is then infinitely thin.

Consider a cylindrical, inhomogeneous plasma with a uniform axial field $\mathbf{B} = \hat{\mathbf{z}}\,B$. Using the same basic equations as those in Chapter 10, it is easy to show that

$$i\omega\mu_o j_r = A(E_r + i\Omega E_\theta) \tag{14.33}$$

$$i\omega\mu_o j_\theta = A(E_\theta - i\Omega E_r) \tag{14.34}$$

$$E_z = 0 \qquad \Omega = \frac{\omega}{\omega_{ci}} \qquad A = \frac{\omega^2}{v_A{}^2(1 - \Omega^2)}$$

where v_A is a function of r. The torsional wave develops a finite b_z component when Ω is finite, so there are no solutions of the type given in Sect. 14.2 with $b_z = 0$. Consequently, k_z is assumed to be independent of r in this Section. For small perturbations of the form

$$f(r)\, e^{i(k_z z + m\theta - \omega t)}$$

we find from Maxwell's equations that

$$\omega b_r = -k_z E_\theta \qquad \omega b_\theta = k_z E_r \tag{14.35}$$

$$i\omega b_z = \frac{1}{r}\frac{\partial}{\partial r}(rE_\theta) - \frac{im}{r} E_r \tag{14.36}$$

$$\mu_o j_r = \frac{im}{r} b_z - ik_z b_\theta \tag{14.37}$$

$$\mu_o j_\theta = ik_z b_r - \frac{\partial b_z}{\partial r} \tag{14.38}$$

$$\nabla \cdot \mathbf{b} = \frac{1}{r}\frac{\partial}{\partial r}(rb_r) + \frac{im}{r} b_\theta + ik_z b_z = 0 \tag{14.39}$$

Elimination of j_r and j_θ from (14.33) to (14.38) gives

$$b_r = \frac{ik_z}{k_\perp{}^2}\left[\frac{\partial b_z}{\partial r} + \frac{mG}{rF} b_z\right] \tag{14.40}$$

$$b_\theta = -\frac{k_z}{k_\perp{}^2}\left[\frac{m}{r} b_z + \frac{G}{F}\frac{\partial b_z}{\partial r}\right] \tag{14.41}$$

$$E_r = \frac{i}{V}\left[m \frac{\partial}{\partial r}(rE_\theta) - Gr^2E_\theta \right] \tag{14.42}$$

$$\frac{\partial}{\partial r}\left[\frac{1}{r}\frac{\partial}{\partial r}(rE_\theta)\right] + FE_\theta = iGE_r + im\frac{\partial}{\partial r}\left[\frac{E_r}{r}\right] \tag{14.43}$$

where $\quad F = A - k_z{}^2 \qquad G = \Omega A$

$$k_\perp{}^2 = \frac{F^2 - G^2}{F} \qquad V = Fr^2 - m^2$$

$k_\perp{}^2$ is a function of r since F, G and A are all functions of r. When m = 0 we find from (14.39) to (14.41) that

$$\frac{\partial}{\partial r}\left[\frac{1}{r}\frac{\partial}{\partial r}(rb_r)\right] = - k_\perp{}^2 b_r \tag{14.44}$$

When $m \neq 0$, it is easier to work with the **E** field components, eliminating E_r from equations (14.42) and (14.43) to give

$$\frac{\partial}{\partial r}\left[Y \frac{\partial E_\theta}{\partial r}\right] = (mX - Z)E_\theta \tag{14.45}$$

where $\quad Y = \dfrac{Fr^3}{V} \qquad X = \dfrac{\partial}{\partial r}\left[\dfrac{U}{V}\right]$

$$U = Gr^2 - m \qquad Z = rF - \frac{1}{r} - \frac{U^2}{rV}$$

Eqs. (14.44) and (14.45) are both singular, when $\Omega < 1$, at the Alfven resonance layer where F = 0 and $k_\perp{}^2 = \infty$. There is also a singularity in Y in (14.45) at the layer where V = 0. There is no singularity in the wave fields when V = 0 but the singularity in Y can present a numerical problem. This problem can be avoided by multiplying both sides of (14.45) by V^2 before integration since $\partial Y/\partial r$ contains a term proportional to $1/V^2$.

Solutions of (14.44) and (14.45) are given in Fig. 14.7. These solutions were obtained numerically for a cylindrical hydrogen plasma having a density profile of the form $n_e = n_o(1 - 0.95\ r^2/a^2)$ and a conducting wall at a = 0.2 m. The boundary condition at the wall is $E_\theta(a) = 0$ or, from equation (14.35), $b_r(a) = 0$.

Comparison of the results in Fig. 14.7 with those for a homogeneous plasma (Fig. 9.4b) shows that the compressional wave modes are not strongly affected by the density gradient. The dispersion relations and cutoff frequencies are very similar to those for a homogeneous plasma having the same average density as the inhomogeneous plasma. The dashed section of the m = +1 mode in Fig. 14.7 intersects the Alfven continuum. The wave fields are singular in this region due to an Alfven resonance near the plasma edge.

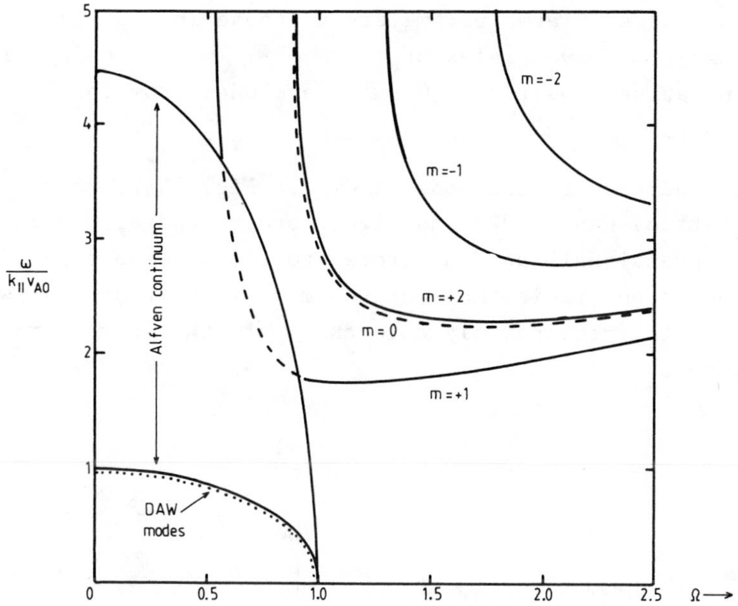

Fig. 14.7 Dispersion relations for the first radial mode Alfven waves in a hydrogen plasma with a parabolic density profile, $n_o = 4 \times 10^{19}\ m^{-3}$. The phase velocity is normalised to v_{A0}, the Alfven speed at r = 0.

The Alfven continuum shown in Fig. 14.7 is bounded by the $k_\perp^2 = \infty$ locus at r = 0 (where $n_e = 4 \times 10^{19}\ m^{-3}$) and the $k_\perp^2 = \infty$ locus at r = a (where $n_e = 0.2 \times 10^{19}\ m^{-3}$). Within the continuum there is a continuous range of solutions where the wave fields are singular at a resonance surface located in the range $0 < r < a$.

The dispersion relation for discrete Alfven wave (DAW) modes lies slightly below the lower boundary of the Alfven continuum, as shown by the dotted curve in Fig. 14.7. Typical b_r profiles for these modes are shown in Fig. 14.8a. The most notable feature is that the m > 0 wave fields are very localised near r = 0. This is due to the fact that m > 0 modes are right hand polarised in the central region of the plasma and hence this region shrinks as the wave frequency is increased, as described in Section 10.3. Similar effects occur for the compressional wave fields. For example, Fig. 14.8b shows the b_r profiles for the first radial mode m = 0, +1 and −1 compressional waves at $\Omega = 1.3$ for the same plasma parameters as those in Fig. 14.7. In this case, the wave fields of the m = −1 mode, being left hand polarised near r = 0, are excluded from this region when $\Omega \sim 1$.

In addition to the modes shown in Fig. 14.7, there also exist surface wave modes and higher order radial modes of the compressional and discrete torsional wave types (E7, H2). The second radial mode of the m = +1 compressional wave has a cutoff frequency close to that of the first radial m = −1 mode.

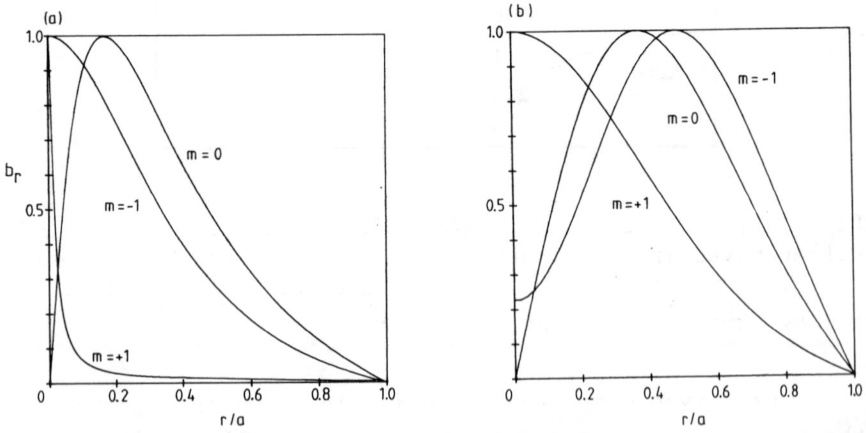

Fig. 14.8 b_r profiles for the first radial m = 0,−1, +1 modes in a hydrogen plasma with a parabolic density profile (same conditions as Fig. 14.7). (a) DAW modes at $\Omega = 0.5$ (b) compressional waves at $\Omega = 1.3$.

14.8 Behavior near an isolated cutoff

The behaviour of Alfven waves in an inhomogeneous plasma is strongly influenced by the existence of resonance ($k_\perp^2 = \infty$) and cutoff ($k_\perp^2 = 0$) surfaces in the plasma. A wave propagating towards a resonance undergoes mode conversion, as described in Section 14.4. The effect of a cutoff is incorporated in the numerical solutions of Section 14.7, but the physics is not at all obvious. To make it obvious, we can model wave behaviour near a cutoff by considering a hypothetical situation where

$$\frac{\partial^2 b}{\partial x^2} = \frac{1}{v^2}\frac{\partial^2 b}{\partial t^2} = -\frac{\omega^2}{v^2} b = -k_x^2 b \qquad (14.46)$$

and where k_x is a function of x. Assuming for example that $k_x^2 = x$, then the solution of (14.46) can be expressed in terms of Airy or Bessel functions,

$$b(x) = b_o \, Ai(-x) = \frac{b_o}{\pi}\int_0^\infty \cos\left[\frac{t^3}{3} - xt\right] dt$$

$$= \frac{b_o\sqrt{x}}{3}\left[J_{-\frac{1}{3}}\left(\frac{2x^{3/2}}{3}\right) + J_{\frac{1}{3}}\left(\frac{2x^{3/2}}{3}\right)\right]$$

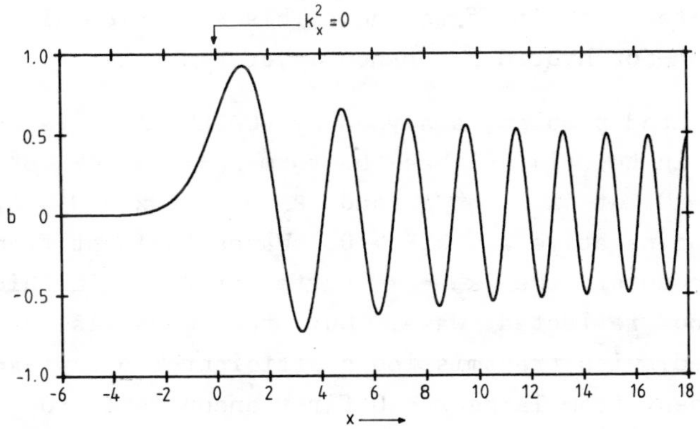

Fig. 14.9 Wave fields near an isolated cutoff

Alternatively, it is easy to solve Eq.(14.46) numerically, for example by the finite difference method described in Chapter 16. The solution is given in Fig. 14.9. The main features of this solution are (a) when $x > 0$, λ decreases as x increases and approaches $2\pi/k_x$ at large x, (b) when $x < 0$, the wave is evanescent, with $k_x^2 < 0$ and (c) the solution for $x > 0$ represents a standing wave. There are no propagating solutions of (14.46). We conclude that a wave incident on a $k_\perp^2 = 0$ cutoff is totally reflected. The wave fields on the far side of the cutoff are not zero. As a result, wave energy can tunnel through a cutoff and emerge as a propagating wave if $k_\perp^2$ changes from a negative back to a positive value, as illustrated in Fig. 11.4. This effect is described in the next Section.

14.9 Behaviour near a combined resonance and cutoff

A wave propagating towards a region where $k_\perp \to \infty$ will be partially reflected, partially transmitted and partially absorbed. An analytical solution for the wave fields in the immediate vicinity of an Alfven resonance layer was given in Sect. 14.4. In that case, the resonance layer was infinitely thin, since the equations were derived using the ideal MHD approximation. In this Section, we consider the situation shown in Fig. 11.4 where a $k_\perp = \infty$ layer is located between two $k_\perp = 0$ cutoff layers, the distance between each layer being finite at finite frequency. This situation also occurs near an ion-ion hybrid resonance layer.

A related problem, analysed by BUDDEN (A3), is the case where k_x^2 in Eq. (14.46) has the form $k_x^2 = \beta/x + \beta^2/\eta^2$ so that $k_x = 0$ at $x = -\eta^2/\beta$ and $k_x = \infty$ at $x = 0$. At large positive or negative x, $k_x^2 > 0$. Waves incident from large $x > 0$ encounter the $k_x = \infty$ surface first. In this case, there is no reflected wave, but the wave is partially transmitted, with transmission coefficient $T = \exp(-\pi\eta/2)$. A wave incident from large $x < 0$ first encounters the $k_x = 0$ cutoff. In this case, $R = 1 - \exp(-\pi\eta)$ and $T = \exp(-\pi\eta/2)$ where R is the reflection coefficient. In both cases, $R^2 + T^2 < 1$, indicating that wave energy is absorbed as a result of the resonance. A more realistic, warm plasma model shows that wave energy is mode-converted to a small λ_x wave type (a kinetic Alfven or ion Bernstein wave) and that this wave is subsequently damped. Note that at large η, corresponding to large separation between the $k_x = 0$ and $k_x = \infty$ surfaces, an incident wave is either totally reflected or totally absorbed, depending on whether it propagates in the +x or −x direction.

Now suppose that $k_x{}^2$ in (14.46) is given by

$$k_x{}^2 = A/(x + i\epsilon) - Bx \tag{14.47}$$

where A and B are constants and where ϵ is a small, artificial damping term. Eqs. (14.46) and (14.47) provide a simplified model of (14.44) or (14.45). The variation of $k_x{}^2$ with x when $\epsilon = 0$ is shown in Fig. 14.10.

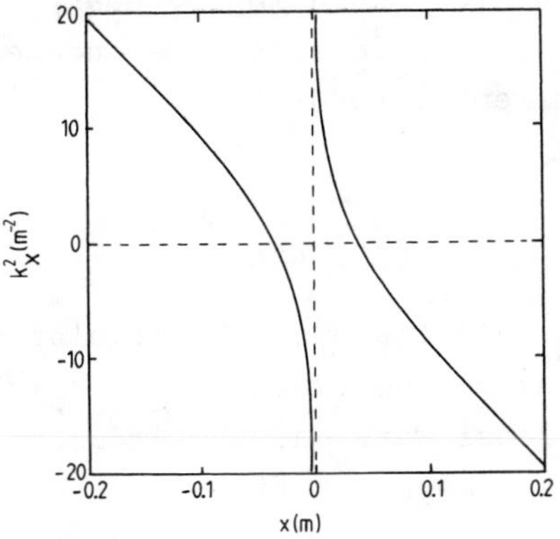

Fig. 14.10 $k_x{}^2$ vs x with A = 0.1 and B = 100

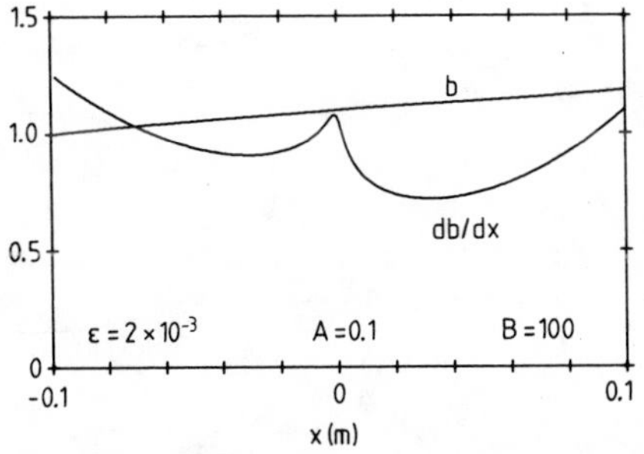

Fig. 14.11 Solutions of Eq.(14.46) when $k_x{}^2$ varies as shown in Fig. 14.10.

When $\epsilon \neq 0$, Eq. (14.46) can be integrated numerically by allowing b to have real and imaginary parts, $b = b_r + ib_i$. Equating the real and imaginary parts of (14.46) yields two second order differential equations which can be solved simulaneously by the finite difference scheme described in Chapter 16. Typical results are shown in Fig. 14.11.

At first sight, the results in Fig. 14.11 are surprising since there is no sign of a resonance in the b field at $x = 0$. The resonance is only evident in db/dx, as shown in Fig. 14.11. These results are similar to those found in Section 14.4 and can be interpreted to mean that wave energy is mode converted near $x = 0$ to a wave with a large $b_x \sim db_z/dx$ component. A full discussion of the power dissipation near the resonance layer is given by STIX (A8, Ch.10). However, these results are mainly of qualitative interest. Full wave solutions, incorporating boundary conditions and specific antenna configurations are needed in order to compare with experimental results.

SPACE FILLER

The Earth's dipole field can be simulated in cylindrical coordinates using a model where $\mathbf{B} = \hat{\theta}\, B_0/r$ and $\rho = \rho_0/r^3$. Show that the torsional wave has the property that $b_\theta = 0$, where b_θ is the wave magnetic field in the θ direction, and that the Poynting vector is in the θ direction. A similar model, with $\rho = \rho_0$, can be used to describe a rectangular cross section tokamak constructed from concentric cylinders and two annular lids.

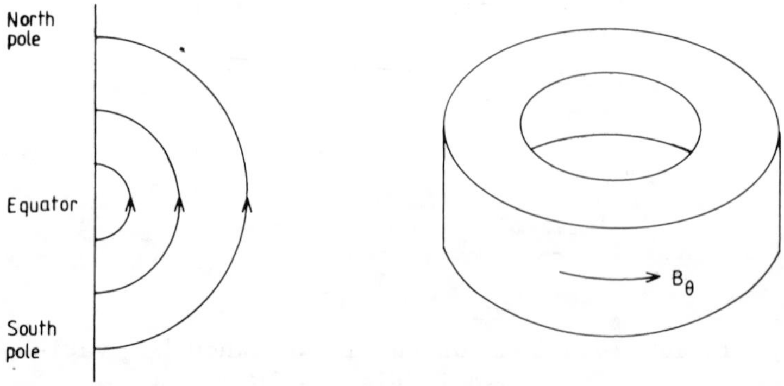

15. RESISTIVE DAMPING

There are several mechanisms available to convert Alfven wave energy into particle energy, the most important being Landau damping, ion cyclotron damping and collisional (viscous and resistive) damping. Frequently, wave damping may involve several different processes. In this chapter we consider purely resistive damping of Alfven waves, arising from collisions between electrons and ions, using a single fluid model of the plasma. The discussion is further simplified by considering low wave frequencies where Ohm's law has the form

$$\mathbf{E} + \mathbf{v} \times \mathbf{B} = \eta \mathbf{j} \tag{15.1}$$

where
$$\eta = 65.3\ Z\ \frac{\ell n\ \Lambda}{T_e^{1.5}}$$

is the plasma resistivity, Z the ion charge, T_e the electron temperature in degrees Kelvin, and $\ell n\ \Lambda$ is the Coulomb logarithm. For laboratory plasmas, $\ell n\ \Lambda \sim 10$. For example, when $Z = 1$, $T_e = 10^6$ K, $\eta = 6.5 \times 10^{-7}\ \Omega$-m. By comparison, the resistivity of copper is $0.17 \times 10^{-7}\ \Omega$-m at 20°C. The plasma temperature has to exceed 1 keV before it becomes a better conductor than copper.

In a metallic conductor, electromagnetic waves penetrate a distance of only a few skin depths through the conductor. The skin depth is small in good conductors and large in poor conductors. The opposite effect is normally observed for Alfven waves in a magnetised plasma; wave damping is normally reduced as the resistivity is reduced. In fact, it was generally thought, up until recently, that resistive damping would be absent in a plasma of zero resistivity. In 1984–5, several papers suddenly appeared in the literature showing that this is not always the case (H6,H7,H10,H12,H15).

The rate of ohmic dissipation is given by ηj^2, in watts/m^3. In a metallic conductor, $j = E/\eta \to \infty$ as $\eta \to 0$. For a given E field, the rate of ohmic dissipation, or the rate at which wave energy is damped, therefore increases as η is reduced. This is not generally true in a plasma, since j usually remains finite when $\eta \to 0$. It is not obvious, at least from (15.1), that j should remain finite as $\eta \to 0$, but Maxwell's $\nabla \times \mathbf{b} = \mu_o \mathbf{j}$ equation generally ensures that it does. There are some circumstances, however, where j_z may approach infinity in a plasma as $\eta \to 0$. This effect is described in detail in Section 15.2. It turns out that resistive damping of the torsional wave may persist even in the limit $\eta \to 0$.

Equation (15.1) can be combined with the equation of motion $\rho\, \partial \mathbf{v}/\partial t = \mathbf{j} \times \mathbf{B}$ to give, with $\mathbf{B} = \hat{\mathbf{z}}\, B$,

$$\mu_o j_x = -\frac{i\omega E_x}{(v_A{}^2 - iR)} \qquad \mu_o j_y = -\frac{i\omega E_y}{(v_A{}^2 - iR)} \tag{15.2}$$

and

$$\mu_o j_z = \omega E_z / R \tag{15.3}$$

where $\quad v_A{}^2 = B^2/(\mu_o \rho) \quad$ and $\quad R = \omega\eta/\mu_o$.

Now consider a homogeneous plasma and assume perturbations of the form $\exp[i(k_x x + k_y y + k_z z - \omega t)]$. The components of Maxwell's curl equations are:

$$\left.\begin{aligned} \omega b_x &= k_y E_z - k_z E_y \qquad & \mu_o j_x &= i(k_y b_z - k_z b_y) \\ \omega b_y &= k_z E_x - k_x E_z \qquad & \mu_o j_y &= i(k_z b_x - k_x b_z) \\ \omega b_z &= k_x E_y - k_y E_x \qquad & \mu_o j_z &= i(k_x b_y - k_y b_x) \end{aligned}\right\} \tag{15.4}$$

The main difference between these equations and those for a plasma of zero resistivity, equation (3.6), is that E_z is not zero in (15.4). Combining the set (15.2) to (15.4) gives

$$\left.\begin{aligned} (\omega^2 + iRk_y{}^2 - Vk_z{}^2)b_x &= iRk_x k_y b_y - Vk_x k_z b_z \\ (\omega^2 + iRk_x{}^2 - Vk_z{}^2)b_y &= iRk_x k_y b_x - Vk_y k_z b_z \\ (\omega^2 - Vk_x{}^2 - Vk_y{}^2)b_z &= -Vk_z(k_x b_x + k_y b_y) \end{aligned}\right\} \tag{15.5}$$

where $\qquad V = v_A{}^2 - iR$.

Since $$\nabla \cdot \mathbf{b} = i(k_x b_x + k_y b_y + k_z b_z) = 0 \tag{15.6}$$

the third expression in equation (15.5) reduces to

$$\omega^2 = (v_A{}^2 - iR)(k_x{}^2 + k_y{}^2 + k_z{}^2) \tag{15.7}$$

provided $b_z \neq 0$. This is the dispersion relation for the compressional wave. The dispersion relation for shear Alfven waves in a resistive plasma is obtained by setting $b_z = 0$ in equation (15.5), which gives two solutions,

$$\omega^2 = k_z{}^2 v_A{}^2 - iR(k_x{}^2 + k_y{}^2 + k_z{}^2) \tag{15.8}$$

and $\omega^2 = k_z{}^2 v_A{}^2 - iRk_z{}^2$. The second solution is simply a special case of equation (15.8) where $k_x = \pm\, ik_y$ and hence $k_\perp{}^2 = k_x{}^2 + k_y{}^2 = 0$. In this case, $b_x = \pm\, ib_y$ and hence the wave is circularly polarized. It is an interesting case since both $k_x{}^2$ and $k_y{}^2$ can be large but there is only light wave damping. In general, resistive damping increases as $k_\perp$ increases.

The imaginary part of (15.7) or (15.8) can be interpreted in several different ways depending on experimental conditions. If ω is real, then at least one of the k's is complex and the wave will attenuate with distance. For plane waves in an infinite, homogeneous plasma, we can let $k^2 = k_x{}^2 + k_y{}^2 + k_z{}^2 = (k_r + ik_i)^2$ to find k_r and k_i. However, if the plasma is bounded in the x and y directions, then standing waves will form in these directions with real k_x and k_y. In this case, k_z is the only component with an imaginary part. If the plasma is inhomogeneous in the x direction, then k_x also develops an imaginary component, as described in Section 15.2.

A second interpretation is that all k's are real and ω is complex. This situation can arise in a three dimensional cavity. An initial perturbation will generate cavity modes which damp exponentially with time, like the transient response of an organ pipe. The real part of ω determines the oscillation frequency and the imaginary part determines the rate at which the amplitude decreases with time. If the cavity is driven at a resonance frequency, then steady state conditions are achieved when the energy dissipated per cycle is exactly replaced by the external source during each cycle. In this case, the quality factor, Q, of the cavity is given by $Q = \omega_r/\omega_i$ where ω_r and ω_i are the real and imaginary parts of ω in the dispersion relation.

15.1 Attenuation of plane waves

Consider plane wave propagation in a homogeneous plasma at an angle θ to **B** where $k_\perp = k \sin\theta$ and $k_\parallel = k \cos\theta$. For torsional waves,

$$(\omega/k)^2 = v_A{}^2 \cos^2\theta - iR \tag{15.9}$$

Now assume that ω and θ are real and $k = k_r + ik_i$. By equating the real and imaginary parts of (15.9), we find that

$$k_r{}^2 = (\omega^2/2q)(v_A{}^2 \cos^2\theta + q^{1/2}) \tag{15.10}$$

$$k_i = \omega^3\eta\,/(2\mu_0 q k_r) \tag{15.11}$$

where $$q = R^2 + (v_A\cos\theta)^4$$

For light damping, where $R \ll v_A{}^2 \cos^2\theta$,

$$\omega/k_r = v_A \cos\theta \quad \text{and} \quad k_i = \omega^2\eta\,/(2\mu_0 v_A{}^3\cos^3\theta)$$

In the opposite limit, $R \gg v_A{}^2 \cos^2\theta$ (eg when $\theta \to 90°$)

$$k_r = k_i = [\mu_0\omega/2\eta]^{1/2} = 1/\delta$$

where δ is the classical skin depth. This result shows that for wave propagation at $\theta = 90°$, the plasma is equivalent to a metallic conductor. At f = 1 MHz, $T_e = 10$ eV, $\eta = 2 \times 10^{-5}$ Ω-m, $\delta = 2.2$ mm. For the same parameters, but at $\theta = 0°$, and $v_A = 5 \times 10^6$m/s, $k_i = 2.5 \times 10^{-6}\text{m}^{-1}$. The attentuation length $L = 1/k_i = 4 \times 10^5$ m.

Clearly, wave attenuation for propagation across B is very strong, while attenuation parallel to B is very weak, at least for the torsional wave. However, a magnetically guided ray, propagating parallel to B, will contain plane wave components propagating almost perpendicular to B. These high $k_\perp$ components attenuate more rapidly, with the result that the ray profile broadens in a direction perpendicular to B as it travels along B.

In the ideal MHD limit, torsional Alfven rays are perfectly guided by the steady magnetic field. Finite resistivity leads to diffusion of the wave fields across the steady field as the ray propagates along the field lines. Experimental results are shown in Fig. 6.1 and a theoretical interpretation of these results is given by SY (G10). Additional calculations are given by BORG et al., (G1). The diffusion of wave energy across field lines is due to the fact that, when η is finite, there is a finite component of the group velocity vector perpendicular to B. Consequently, the wave can reflect off the plasma boundary to form a discrete (waveguide) mode. The discrete modes are described in Section 15.2. However, it is much easier to observe the guided component of the wave energy emerging from an antenna, since it is concentrated within a small part of the plasma cross – section, on the same field lines as the antenna. By the time that energy reaches the boundary, the wave has propagated a large distance from the antenna, and most of the energy will be dissipated resistively in the plasma. Energy which does reflect off the boundary is distributed over the entire plasma cross – section and is therefore difficult to detect due to the small wave amplitude. An equivalent picture emerges if one sums over the infinite set of discrete modes which can propagate in the plasma. The high $k_\perp$ modes constitute the guided ray while the low order (low $k_\perp$ or low m) modes are much smaller in amplitude, especially if they are generated by physically small antennas.

15.2 **The resistive spectrum**

We now consider the discrete spectrum of m = 0 torsional wave modes which can propagate in a cylindrical, inhomogeneous plasma of finite resistivity. When the resistivity is zero, each annular element of plasma behaves independently of its neighbour, as described in Chapter 14. The effect of finite resistivity is to couple adjacent annuli and to introduce wave damping. Although wave damping is to be expected when the plasma has finite resistivity, the amount of damping one calculates is totally unexpected. Solutions obtained in the limit $\eta \to 0$ are NOT the same as those with $\eta = 0$. In fact, the damping rate of a mode localised in any given annular region of the plasma is independent of η. As η is reduced, the wave fields become more localised and more oscillatory in each annular region but the damping rate is unaffected. However, if one examines any particular radial mode, its damping rate approaches zero as $\eta \to 0$ and the wave fields of that mode tend to be localised either at the edge of the plasma or in the centre of the plasma (at r = 0).

To investigate these phenomena, we will consider a cylindrical plasma with a uniform axial field $\mathbf{B} = \hat{\mathbf{z}}\, B$ and a perfectly conducting wall at r = a. It is assumed that the resistivity is independent of radius and that the density profile has the form $n_e = n_0(1 - pr^2/a^2)$ where p is a parameter used to specify the edge density. The value of p is not critical but we will consider solutions with p = 0.75 so that the local Alfven speed varies by a factor of two between r = 0 and r = a. For simplicity, we assume that there is no plasma current.

At low wave frequencies, where Ohm's law is given by $\mathbf{E} + \mathbf{v} \times \mathbf{B} = \eta \mathbf{j}$, the m = 0 torsional and compressional wave types are uncoupled. The b_r and b_z components for the torsional wave are then zero, and it is easy to show that the E_z component satisfies the equation

$$\frac{1}{r}\frac{\partial}{\partial r}\left[\frac{r}{Q}\frac{\partial E_z}{\partial r}\right] = - E_z \tag{15.12}$$

where

$$Q = \frac{i\omega\mu_0}{\eta}\left[1 - \frac{k_z{}^2 v_A{}^2}{\omega^2}\right] - k_z{}^2 \tag{15.13}$$

and v_A is the local Alfven speed. The axial wave number, k_z, is assumed to be independent of r. The quantity, Q, is equal to $k_\perp{}^2$ for a homogeneous plasma, but varies with r in an inhomogeneous plasma. Since Q has an imaginary component, the E_z field in Eq. (15.12) can be expressed in the form

$E_z = E_r + iE_i$ where E_r denotes the real part of E_z (and not the radial component of **E**). Let $1/Q = Y_r - iY_i$. Equating the real and imaginary parts of Eq. (15.12) yields two coupled equations

$$\frac{\partial}{\partial r}\left(r\left[Y_r\frac{\partial E_r}{\partial r} + Y_i\frac{\partial E_i}{\partial r}\right]\right) = -rE_r \tag{15.14}$$

$$\frac{\partial}{\partial r}\left(r\left[Y_r\frac{\partial E_i}{\partial r} - Y_i\frac{\partial E_r}{\partial r}\right]\right) = -rE_i \tag{15.15}$$

We can now assume either that k_z is real and ω is complex or that ω is real and k_z is complex. Both approaches yield equivalent results, as summarised in Figs. 15.1 and 15.5. Solutions of Eqs. (15.14) and (15.15) for real ω and complex k_z are given in Fig. 15.1. These solutions were obtained, using the finite difference technique described in Chapter 16, by searching for values of $k_r = \mathrm{Re}(k_z)$ and $k_i = \mathrm{Im}(k_z)$ to satisfy the boundary conditions $E_r(a) = 0$ and $E_i(a) = 0$. The finite difference forms of Eqs. (15.14) and (15.15) can easily be solved simultaneously to yield expressions for E_r and E_i at each new integration step. The solutions were obtained by starting with an arbitrary value of E_r at $r = 0$. The starting value of E_i at $r = 0$ can also be chosen arbitrarily (e.g. $E_i = 0$ or $E_i = 1$) since there is an arbitrary phase difference between E_r and E_i at $r = 0$. Solutions were obtained by fixing k_i and varying k_r until $E_r(a) = 0$ and then repeating this procedure for different values of k_i so that both $E_r(a) = 0$ and $E_i(a) = 0$.

As indicated in Fig. 15.1, there is an infinite set of solutions at any given value of η, each solution corresponding (with a few exceptions described below) to a different radial mode. Each radial mode has a different number of zero crossings, of the E_z field, between $r = 0$ and $r = a$, and each mode has a different set of k_r, k_i coordinates. The variation of $E_r = \mathrm{Re}(E_z)$ with r is shown in Fig. 15.2 for several different modes.

The most surprising result of these calculations is that the solutions for different value of η all lie on the same, fixed trajectories in the k_r, k_i plane. Intuitively, one would expect the solutions to converge onto the real axis as $\eta \to 0$ since there is no wave damping when $\eta = 0$. When $\eta = 0$, there is a continuum of solutions with $k_z(r) = \omega/v_A(r)$ corresponding to the continuous variation in the local Alfven speed betwen r = 0 and r = a.

Another surprising feature of the solutions is that there are three distinct trajectories, all meeting at a common bifurcation point ($k_r \sim 1.1 m^{-1}$, $k_i \sim 0.2 m^{-1}$ for the parameters of Fig. 15.1). Two of the trajectories leave this point and connect to the ends of the ideal continuum while the other trajectory disappears off the top edge of Fig.15.1 and asymptotically approaches the straight line $k_i = k_r$.

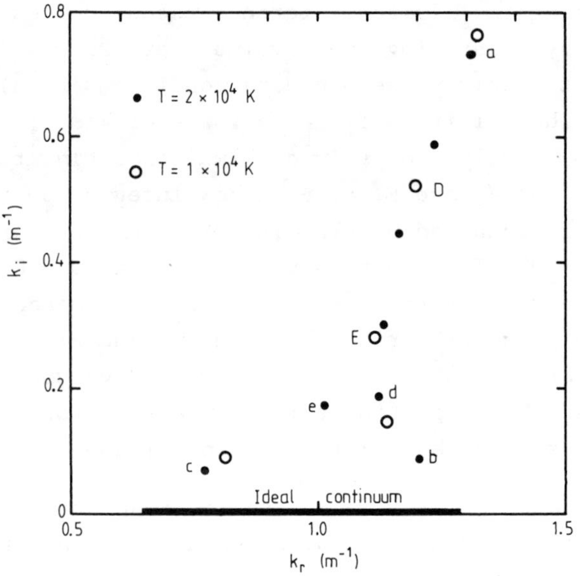

Fig. 15.1 The complex k_z spectrum for m = 0 torsional wave modes in a resistive plasma. Field profiles for modes a,b and c are given in Fig. 15.2. Modes D and E (at $T = 1 \times 10^4$ K) evolve into modes d and e respectively at $T = 2 \times 10^4$ K. Parameters are B = 1.0 T, a = 0.1m , f = 1.0 MHz, $n_e = 2 \times 10^{19}(1-0.75r^2/a^2)\ m^{-3}$, for hydrogen at $T = 1 \times 10^4$ K or $T = 2 \times 10^4$ K.

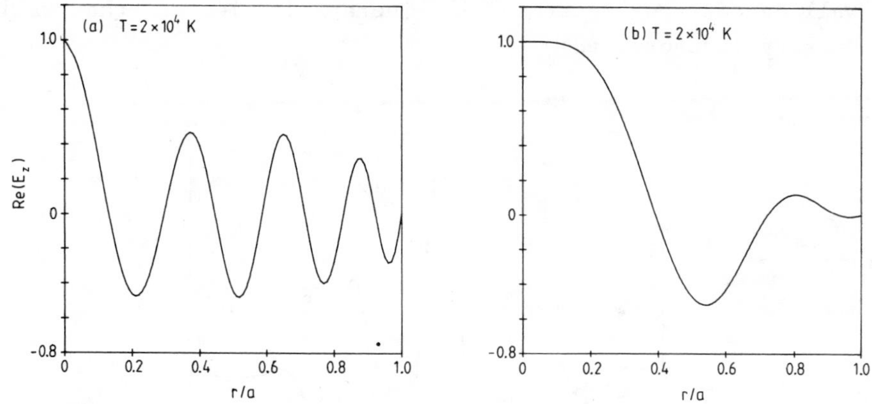

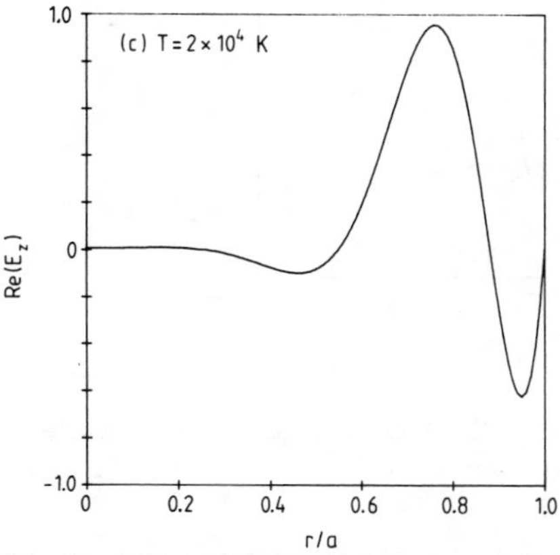

Fig. 15.2 Radial profiles of $Re(E_z)$ for the modes labelled a, b and c in Fig. 15.1.

Each radial mode moves continuously along one of the three trajectories in the k_i, k_r plane, as η is varied. In a very resistive plasma, all modes are heavily damped. The higher order modes (having more zero crossings) lie at successively higher k_r positions along the trajectory $k_i = k_r$. The low order modes lie closer to the bifurcation point, in successive order. As η is reduced, all modes move along the $k_i = k_r$ trajectory, heading for the bifurcation point. When they reach this point, they head along one of the other two trajectories towards the ends of the continuum. However, the number of zero crossings of a given mode can change as it moves along its trajectory, especially near the bifurcation point. As a result, several different modes, with the same number of zero crossings, can appear in different locations in the k_i, k_r plane. The 1st, 2nd and 3rd radial modes do not even appear in Fig. 15.1, although they are present in a homogeneous plasma, as shown in Fig. 15.4. These low order modes continuously deform into higher order modes as the wall density parameter p is increased (ie as the wall density is decreased).

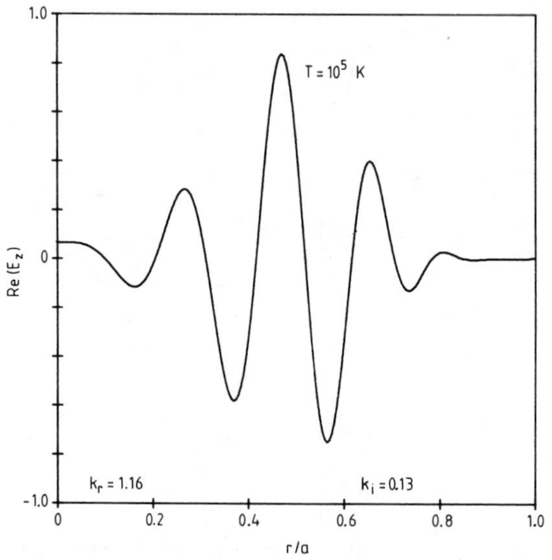

Fig. 15.3 Re(E_z) profile at T = 10^5 K for an m = 0 mode localised near r/a = 0.5.

The net result of decreasing η is that more and more modes appear along any given segment of the solution trajectories and the modes become more oscillatory since the low order modes accumulate near the ends of the continuum. The high order modes become more localised as η is reduced, and peak in amplitude close to the radius where $k_r = \omega/v_A(r)$. For example, Fig. 15.3 shows a typical result at T = 10^5 K for the same parameters as those in Fig. 15.1. The limit $\eta \to 0$ cannot be investigated numerically since there is an infinite number of solutions present for any value of η, regardless of how small one choses η. However, it is comforting to note that all modes head towards the real axis as η is reduced, accumulating at either end of the ideal continuum.

It is interesting to compare these results with those for a homogeneous plasma. When the density is uniform the spectrum is much simpler and can be obtained analytically. In the uniform density case, the dispersion relation, from equation (15.8) or (15.13) is

$$\omega^2 = k_z{}^2 v_A{}^2 - iR(k_\perp{}^2 + k_z{}^2)$$

where $R = \omega\eta/\mu_o$. This relation simplifies, either when $k_\perp{}^2 \gg k_z{}^2$ or when $R \ll v_A{}^2$ to give

$$k_z{}^2 = \frac{\omega^2}{v_A{}^2} + \frac{iRk_\perp{}^2}{v_A{}^2} = (k_r{}^2 - k_i{}^2) + 2ik_r k_i$$

where $k_z = k_r + ik_i$. Equating the real and imaginary parts of this relation, we obtain

$$k_r{}^2 = \frac{\omega^2 + (\omega^4 + R^2 k_\perp{}^4)^{1/2}}{2v_A{}^2} \tag{15.16}$$

and $$k_i = [k_r{}^2 - \omega^2/v_A{}^2]^{1/2} \tag{15.17}$$

For the conditions of Fig. 15.1 ($B = 1T$, $n_e = 10^{19}\ m^{-3}$, $f = 1MHz$, $T = 10^4\ K$), $R/v_A{}^2 = 6.8 \times 10^{-5}$ and hence the solutions (15.16) and (15.17) are of interest. Eq. (15.17) indicates that all solutions in a homogeneous plasma will lie on a curved line in the k_i, k_r plane, intersecting the real axis ($k_i = 0$) at $k_r = \omega/v_A$ and approaching asymptotically the line $k_i = k_r$ at large k_r. For any given value of η, an infinite set of $m = 0$ modes can be found, satisfying the boundary condition $E_z(a) = 0$. For a uniform density plasma, (15.12) reduces to the standard Bessel equation

$$\frac{\partial^2 E_z}{\partial r^2} + \frac{1}{r}\frac{\partial E_z}{\partial r} + k_\perp{}^2 E_z = 0$$

and hence $$E_z = E_o J_o(k_\perp r).$$

Since $E_z(a) = 0$, the allowed values of $k_\perp$ are given by the zeros of the J_o function,

$$k_\perp a = 2.405,\ 5.520,\ 8.654,\ 11.79,\ 14.93,\ \ldots$$

Note that $k_\perp$ is real for a homogeneous plasma and E_z is also real. If one assumes that E_z has an imaginary component, then the boundary conditions can only be satisfied when the imaginary part of E_z is zero. The solutions for a homogeneous plasma are shown in Fig. 15.4, using parameters similar to those in Fig. 15.1 for comparison.

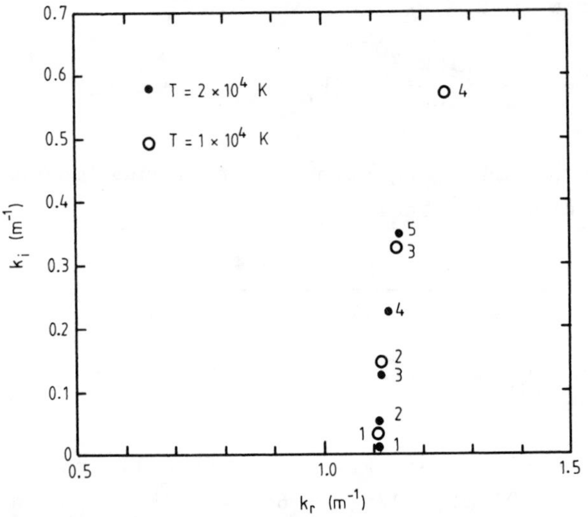

Fig. 15.4 The complex k_z spectrum for a homogeneous hydrogen plasma with B = 1.0 T, f = 1.0 MHz, a = 0.1m, $n_e = 1.5 \times 10^{19}$ m^{-3} at T = 1 × 10^4 K or T = 2 × 10^4 K. The radial mode numbers are indicated on the Figure.

The connection between solutions for a homogeneous plasma (Fig. 15.4) and the inhomogeneous plasma is fairly obvious, especially at large k_i. The reason for the bifurcation in an inhomogeneous plasma is not so obvious but has been analysed by several authors (eg H7) in connection with the complex ω, real k_z spectrum. This spectrum is shown schematically in Fig.15.5. The spectrum has the same qualitative features as the complex k_z spectrum, and retains these features even when $m \neq 0$ and also in a current carrying plasma (H4, 5, 6, 7, 10, 12, 14, 15). However, the analysis of these spectra is rather complicated, even without the inclusion of other important effects such as finite electron mass and temperature.

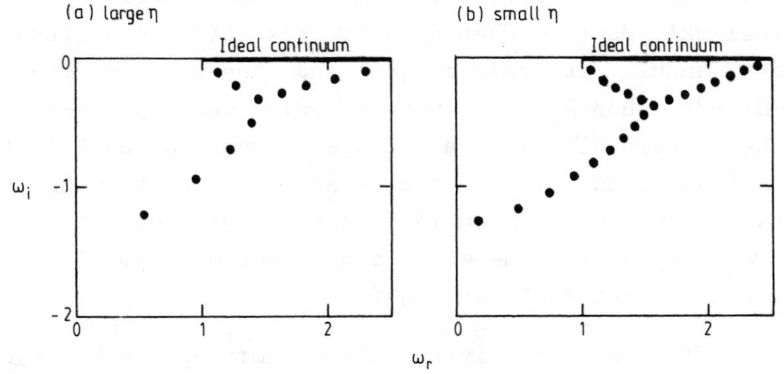

Fig. 15.5 The complex ω spectrum for torsional waves in an inhomogeneous plasma.

15.3 Phase-mixing in a resistive plasma

As described in Chapter 14, an m = 0 torsional wave disturbance has the unique property, in ideal MHD, that it is phase – mixed by a density gradient. In a plasma of zero resistivity, the only wave magnetic field component is b_θ and the only velocity component is v_θ. Adjacent concentric annuli behave independently and the torsional wave propagates at the local Alfven speed in each annulus. Finite resistivity introduces an E_z component, but b_r, b_z, v_r and v_z all remain zero. The Poynting vector therefore develops a component in the radial direction but it remains small compared with the z component for most conditions of interest. Nevertheless, adjacent annuli are no longer totally independent and the phase-mixing process needs to be re-examined.

Phase-mixing in a resistive plasma has been analysed independently by SY (G10) in connection with the results shown in Fig. 6.1 and by NOCERA et al. (M12) in relation to the problem of wave heating in the solar corona. The analysis is surprisingly complicated but the results are easily summarised. The behaviour of an m = 0 wave depends on the magnetic Reynolds number $R_m = \mu_0 \omega a^2/\eta$ where a is the plasma radius. Alternatively, $R_m = 2(a/\delta)^2$ where δ is the classical skin depth. When $R_m \gg 1$, resistive coupling of adjacent annuli is unimportant and the wave is strongly phase-mixed. When $R_m < 1$, there is only weak phase-mixing and the torsional wave no longer travels at exactly the local Alfven speed. These results are consistent with the results given in Section 15.1 where it was found that the phase velocity of a plane wave is unaffected at small η but is strongly affected at large η.

The effect of resistivity on phase-mixing can be assessed, from a different point of view, in terms of the eigenmode solutions given in Section 15.2. Consider a localised source in the plasma (such as the antenna shown in Fig. 6.1) which generates an azimuthally symmetric disturbance. The resulting wave can be regarded as a superposition of the infinite set of discrete modes plotted in Fig. 15.1. Since each mode propagates at a different phase velocity, the resulting wave is phase-mixed. In a plasma of zero resistivity, the wave field gradients, in a direction perpendicular to **B**, would *increase* with distance from the source due to phase-mixing, as shown in Fig. 14.2. However, in a resistive plasma, and for a localised source, the wave field gradients will *decrease* with distance from the source, since the high $k_\perp$ modes will attenuate more rapidly with distance. The radial profile of the ray emerging from the antenna therefore broadens as it propagates away from the source, as observed by CROSS(G3).

For the parameters in Fig. 15.1, the attenuation length of mode "a" is about 1.3m and the attenuation length of modes b and c is about 10m. Consequently, the amplitude of mode "a" will be reduced by a factor of about 2000 over a distance of 10m, while modes b and c decrease in amplitude by a factor of e = 2.7. However, a localised source will generate higher order modes more efficiently than the low order modes, and these are the only modes observed close to the antenna.

Conversely, an extended source will couple more efficiently to the low order modes. The question then arises as to whether these modes will suffer any increased damping as a result of phase-mixing. The short answer must be NO, because each mode is linearly independent of the other and its damping rate is as shown in Fig. 15.1. An even more subtle question arises about the fate of any given mode. Suppose we devise an antenna which launches a pure mode, say mode c in Fig. 15.2. This mode is only weakly damped, with $k_i \ll k_r$. The question to ask about this mode is why it is NOT phase-mixed by the density gradient. Why should each part of the disturbance propagate with the same k_z?

The answer seems to be connected with the fact that the wavelength in the radial direction is shorter near the plasma edge than near the plasma centre. This effect occurs for all the modes plotted in Fig. 15.1. Consequently, the plasma edge region contains a $k_\perp$ spectrum which is richer in high $k_\perp$ components than at the plasma centre. It is easy to show from (15.8) that, for real $k_\perp$ and complex k_z, the phase velocity of a wave at high $k_\perp$ is less than the Alfven speed and decreases as $k_\perp$ decreases. Consequently, it can be argued that the wavefront near the plasma edge does not travel faster than in the high density region since the high $k_\perp$ components hold it back. Alternatively, if we insist that k_z is independent of r, then the radial wavelength near the plasma edge must be shorter than at the plasma centre.

SPACE FILLER

Magnetic field lines in a tokamak lie on concentric surfaces as shown in Fig. 15.6(a) on the next page. The successive points of intersection of a field line in any given cross section of the plasma, after each toroidal trip, are indicated by numbered dots. A field line on a $q = 2$ surface forms a closed loop after two toroidal trips. A field line on a $q = 3$ surface forms a closed loop after 3 toroidal trips. Now suppose that there is a small radial magnetic field perturbation of the form $b_r = b_0 \sin(2\theta)$, generated by a helical current sheet or filament along the field lines near the q = 2 surface. The b_r field will cause a field line to move radially outwards or inwards depending on the sign of b_r. Successive points of intersection of a field line then appear as shown in Fig. 15.6(b). The points of intersection lie on a flux surface known as a magnetic island. The successive displacements of a field line, after each toroidal trip, decrease to zero near the X-points of the island, where $b_r = 0$. b_r is also zero at the O-points (the widest parts of the island).

(i) Given that the strength of a magnetic field is indicated by the number of field lines per unit area, does this mean that the toroidal field increases towards infinity at the X-points?

(ii) Given that electrons follow field lines, will electrons tend to accumulate at the X-points?

(iii) Will an Alfven resonance located near a rational surface create a magnetic island?

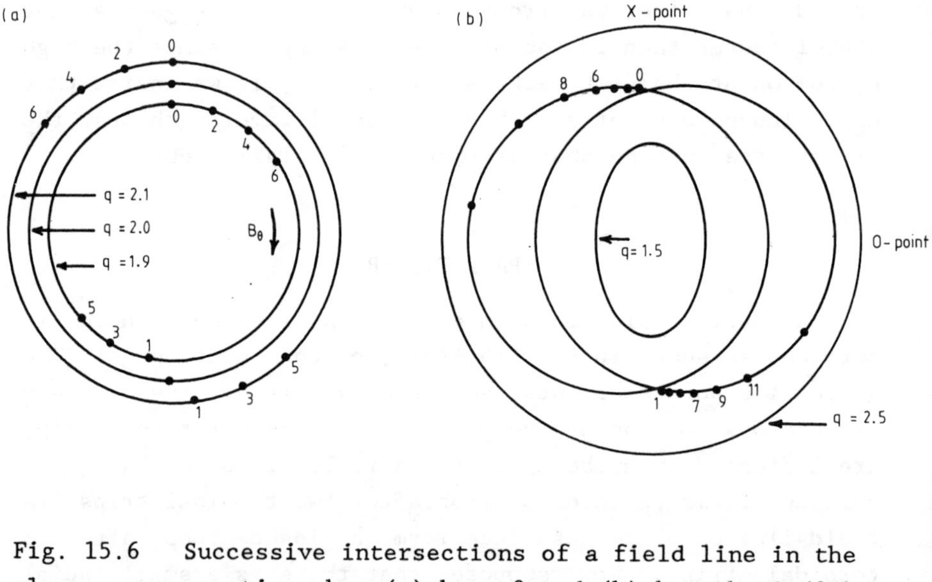

Fig. 15.6 Successive intersections of a field line in the plasma cross section when (a) $b_r = 0$ and (b) $b_r = b_0\sin(2\theta)$.

16. SIMPLE NUMERICAL TECHNIQUES

There are several standard techniques, such as the Runge–Kutta method, available for solving differential equations numerically. A simple method which is easily programmed on a personal computer and which is sufficiently accurate for Bessel type equations is as follows. Many of the equations in this book can be cast in the form

$$\frac{\partial}{\partial r}\left[Y(r)\frac{\partial b}{\partial r} \right] = Z(r)b \qquad (16.1)$$

The variation of b with r can be determined at discrete points separated by a step size $\Delta r = h$ as shown in Fig. 16.1

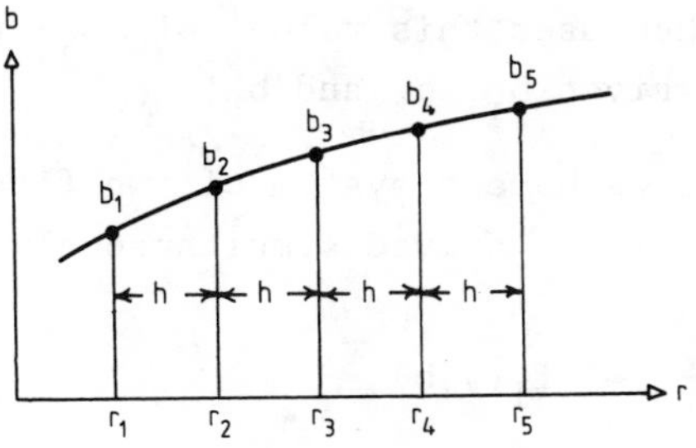

Fig. 16.1 b vs r at discrete points.

The gradients at r_2 and at r_4 are approximately

$$b'(r_2) = (b_3 - b_1)/2h \qquad\qquad b'(r_4) = (b_5 - b_3)/2h$$

$$\therefore \quad \frac{\Delta(Yb')}{\Delta r} = \frac{Y(r_4)b'(r_4) - Y(r_2)b'(r_2)}{2h}$$

$$\therefore \quad Y_4(b_5 - b_3) - Y_2(b_3 - b_1) = 4h^2 Z_3 b_3$$

where $Y_4 = Y(r_4)$ etc.

$$\therefore \quad b_5 = b_3 + \frac{1}{Y_4}\left[Y_2(b_3 - b_1) + 4h^2 Z_3 b_3 \right] \qquad (16.2)$$

Given b_1 and b_3, we can calculate b_5 from Eq. (16.2) since $Y(r)$ and $Z(r)$ are known functions of r. The next value of b is obtained by setting $b_1 = b_3$ and $b_3 = b_5$ and repeating the calculation. To start the integration, one either guesses values of b_1 and b_3 (eg $b_1 = 0$, $b_3 = 0.1$ or $b_1 = b_3 = 1.0$) or uses analytic results. The accuracy of the calculation can be checked by repeating it with a different step size or comparing with known results (solutions of Bessel equations are tabulated in standard texts and described in Sec. 16.1). Usually, sufficient accuracy can be obtained by taking ~100 steps between the centre of the plasma ($r = 0$) and the vessel wall ($r = a$), so a suitable step size is $h \sim a/100$.

The above technique is known as a finite difference method. Another simple technique is the predictor-corrector method. The idea is that one predicts a value for b_5, given b_1 and b_3 and then uses this value of b_5 to get a more accurate value, given b_1, b_3 and b_5'.

Suppose that we have a system of two first order differential equations to be solved simultaneously,

$$\frac{dy}{dr} = f_1(y,b) \tag{16.3}$$

$$\frac{db}{dr} = f_2(y,b) \tag{16.4}$$

For example 16.1) can be written as two first order DE's, $y = Y(r)\partial b$, and $\partial y/\partial r = Z(r)b$. Given y_n and $y_n' = (\partial y/\partial r)_n$ and a step ize $\Delta r = h$, a simple prediction of y_{n+1} is

$$y_{n+1} = y_n + hy_n' \tag{16.5}$$

A better predictor, based on a quadratic fit is

$$y_{n+1} = y_n + \frac{h}{2}(3y_n' - y'_{n-1}) \tag{16.6}$$

For example, consider the equation $dy/dr = 2y$ where $y = e^{2r}$ is the known solution. Let $r_1 = 1$, $h = .5$, $r_2 = 1.5$, $r_3 = 2$. Then

- $y_1 = 7.39$, $y_2 = 20.09$, $y_3 = 54.6$, $y_1' = 14.78$, $y_2' = 40.2$
- Predictor (16.5) gives $y_3 = y_2 + hy_2' = 40.18$.
- Predictor (16.6) gives $y_3 = y_2 + 0.25(3y_2' - y_1') = 46.5$

If the predicted values y_{n+1}, b_{n+1} are substituted into Eqs. (16.3) and (16.4), we obtain values y'_{n+1} and b'_{n+1}. These values can now be used in a quadratic fit corrector, for example

$$y_{n+1} = \frac{1}{3}(4y_n - y_{n-1} + 2hy'_{n+1}) \qquad (16.7)$$

In the above example, $dy/dr = 2y$, we find $y_3' = 2 \times 46.53 = 93.06$ for predictor (16.6) and the corrector then gives $y_3 = 55.34$. Further improvement can be obtained by Newton iteration of the corrector and/or by using a smaller step size. The fastest procedure depends on the equations being solved. The main advantage of the predictor– corrector–iteration technique is that it is possible to take relatively large step sizes even when y is varying rapidly. The reader is referred to standard texts on numerical methods for further details and other methods, but will find Eqs.(16.2) or (16.6) and (16.7) adequate for solving the types of equations described in this book.

16.1 Bessel functions

The general solutions of the Bessel equation

$$\frac{d^2b}{dr^2} + \frac{1}{r}\frac{db}{dr} + \left[k^2 - \frac{m^2}{r^2}\right]b = 0 \qquad (16.8)$$

are given by

$$b = b_1 J_m(kr) + b_2 Y_m(kr) \qquad \text{if} \quad k^2 > 0$$

or

$$b = b_1 K_m(k_o r) + b_2 I_m(k_o r) \qquad \text{if} \quad k_o^2 = -k^2 > 0$$

or

$$b = b_1 r^m + b_2 r^{-m} \qquad \text{if} \quad k^2 = 0$$

where b_1 and b_2 are arbitrary constants, and

$J_m(kr)$ is the Bessel function of the first kind
$Y_m(kr)$ is the Bessel function of the second kind
$K_m(k_o r)$ and $I_m(k_o r)$ are modified Bessel functions.

These functions are plotted in Fig. 16.2 for m = 0,1 and 2. When m is negative,

$$J_{-m}(x) = (-1)^m J_m(x) \qquad Y_{-m}(x) = (-1)^m Y_m(x)$$

$$K_{-m}(x) = K_m(x) \qquad I_{-m}(x) = I_m(x)$$

The radial derivates of the Bessel functions are given by

$$\frac{d}{dr}[J_m(kr)] = kJ_{m-1}(kr) - \frac{m}{r} J_m(kr) = -kJ_{m+1}(kr) + \frac{m}{r} J_m(kr)$$

$$\frac{d}{dr}[I_m(kr)] = kI_{m-1}(kr) - \frac{m}{r} I_m(kr) = kI_{m+1}(kr) + \frac{m}{r} I_m(kr)$$

$$\frac{d}{dr}[K_m(kr)] = -kK_{m-1}(kr) - \frac{m}{r} K_m(kr) = -kK_{m+1}(kr) + \frac{m}{r} K_m(kr)$$

$$J_0'(x) = - J_1(x) \qquad Y_0'(x) = - Y_1(x)$$

When $x \ll 1$

$$J_m(x) = \frac{(x/2)^m}{m!} - \frac{(x/2)^{m+2}}{(m+1)!} + \frac{(x/2)^{m+4}}{2!(m+2)!} - \ldots.$$

$$I_m(x) = \frac{(x/2)^m}{m!} + \frac{(x/2)^{m+2}}{(m+1)!} + \frac{(x/2)^{m+4}}{2!(m+2)!} + \ldots.$$

$$K_0(x) = \frac{x^2}{4} - \left[0.5772 + \ell n_e(x/2)\right] I_0(x) + \ldots \approx - \ell n_e x$$

$$K_1(x) = \frac{1}{x} + I_1(x)\ell n_e(x/2) + \ldots \approx \frac{1}{x}$$

$$K_m(x) \approx \frac{(m-1)!}{2(x/2)^m} \qquad \text{when } m > 0$$

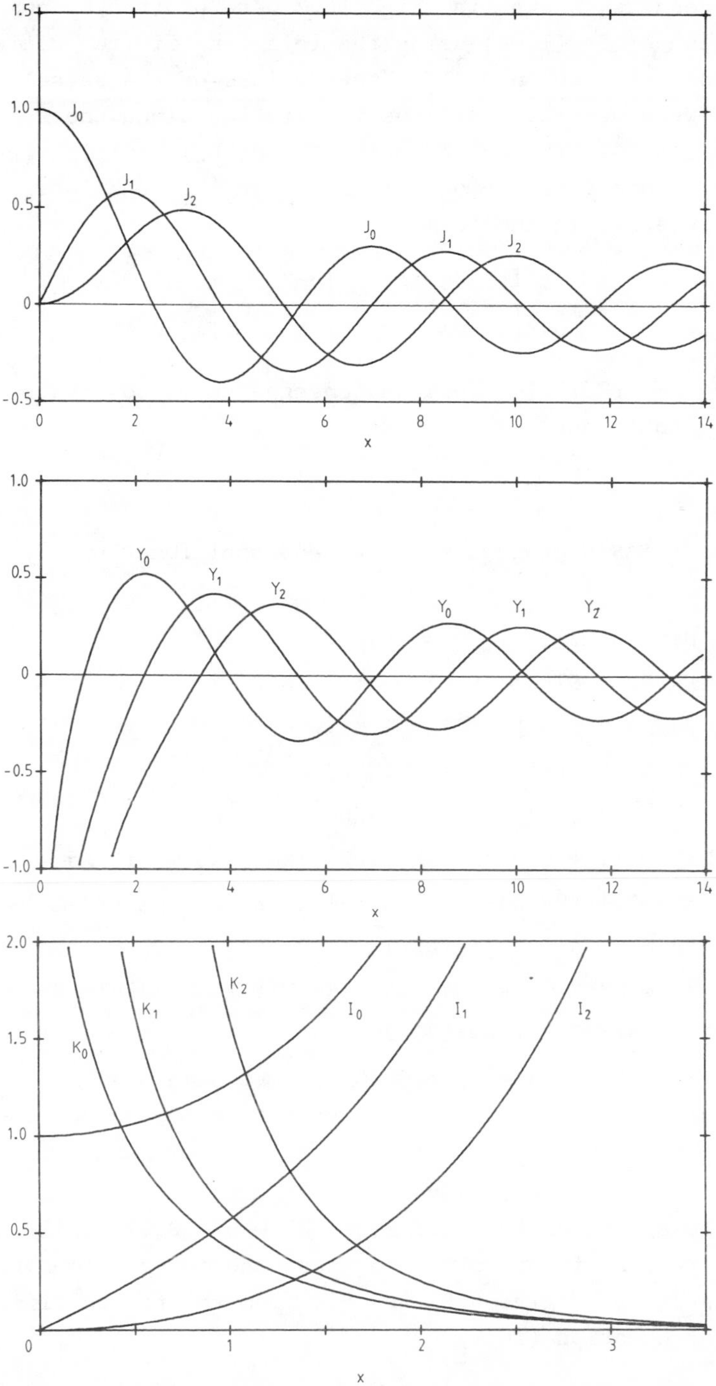

Fig. 16.2 Bessel functions $J_m(x), Y_m(x), K_m(x)$ and $I_m(x)$

The results shown in Fig. 16.2 were plotted by numerical solution of Eq.(16.8) using the following finite difference program, written in BASIC. Known values of the Bessel functions were used to start the integration since the solutions contain an arbitrary constant (b_1 or b_2) multiplying the Bessel functions. To obtain the J_m and Y_m functions, (16.8) was expressed in the form

$$\frac{d}{dx}\left[x \frac{db}{dx} \right] = x \left[\frac{m^2}{x^2} - 1 \right] b \tag{16.9}$$

where x = kr. The −1 term was changed to +1 to obtain the K_m and I_m functions.

BASIC program to compute Bessel functions

```
 10  H1 = 0.02
 20  PRINT " START at x1 = "; : INPUT X1
 30  PRINT " m = "; : INPUT M1
 40  PRINT " b(x1) = "; : INPUT B1
 50  PRINT " b(x3) = "; : INPUT B3
 60  X2 = X1 + H1 :X3 = X2 + H1 :X4 = X3 + H1 :X5 = X4 + H1
 70  B5 =B3+(X2*(B3-B1)+4*H1*H1*X3*B3*(M1*M1/(X3*X3)-1))/X4
 80  PRINT [F8.4 X5]; TAB(12); [D3 B5]
 90  X1 = X3:X2 = X1+H1: X3 = X2+H1: X4 = X3+H1: X5 = X4+H1
100  IF X5 > 14  THEN 120
110  B1 = B3 : B3 = B5 : GOTO 70
120  END
```

Line 10 specifies the step size H1 = h = 0.02 while lines 20–50 request input values to start the integration. Line 70 evaluates B5, given B1 and B3, from the finite difference form of equation (16.9).

17. EXPERIMENTAL TECHNIQUES

At low power levels, Alfven waves can be generated in laboratory plasmas with relatively simple apparatus. The main requirement is an RF source of about 100 W output power (to see signal well above noise) connected via a matching circuit to a loop antenna either surrounding the plasma or located near the plasma edge. In low temperature ($T_e < 10$eV) plasmas, the antenna can be inserted well into the plasma. Alfven waves are easily detected with a small magnetic probe located inside the plasma or near the plasma edge. Alfven wave magnetic fields can also be measured outside the vacuum vessel, even if the vessel is metallic, since the fields penetrate into the laboratory through port openings. Probes inserted into a plasma are normally housed in quartz or pyrex tubes for electrical and thermal insulation. Quartz is better since it has an extremely low coefficient of thermal expansion and therefore survives thermal shock extremely well. Quartz can be heated to red heat and quenched in cold water without cracking. Vycor tubing is cheaper than fused silica, and at 96% SiO_2 is slightly less pure, but works just as well for most applications, even in 100 eV plasmas.

Since the frequency range of interest for most Alfven wave experiments (1 – 30 MHz) covers the amateur radio band, the equipment and techniques available to amateur radio users are eminently suited for Alfven wave experiments. The equipment is readily available at moderate cost, at least at power levels below 2 kW.

17.1 Antennas

Most of the antennas used for experiments in tokamak devices are of all-metal construction to minimise contamination of the plasma. In small, low temperature devices, antennas can be protected with quartz tubing, although unshielded, electrically floating antennas have also been used with success even in tokamak plasmas. More commonly, antennas are constructed from copper, bent into a loop of the desired shape, and electrostatically shielded with strips of stainless steel (or carbonised aluminium to minimise eddy current losses and impurity contamination). The shape and orientation of the loop depends on the wave type being excited and also on the type of wave which needs to be suppressed. Generally, a loop which generates a vacuum magnetic field in say the x direction will couple most efficiently to waves whose main magnetic field component near the antenna is in the x direction. However, the situation can be complicated by wave generation from radial feeder leads and from elements used to support the antenna structure. Antenna design is often a matter of trial and error. Some typical designs are shown schematically in Fig. 17.1.

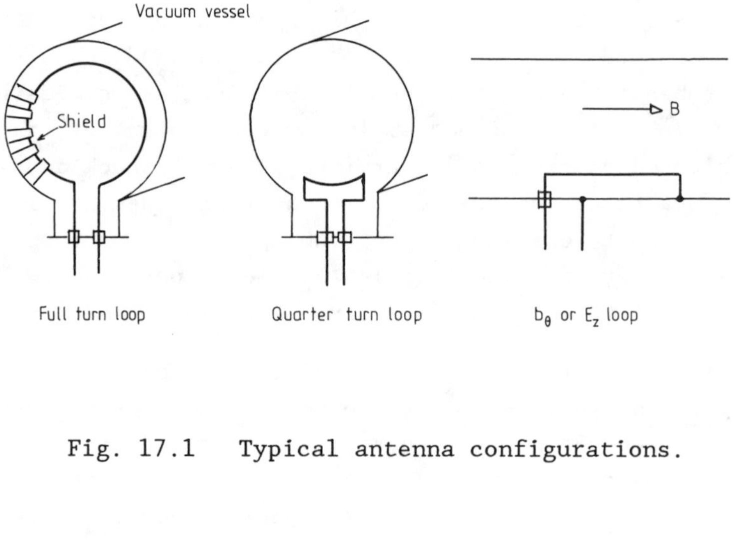

Fig. 17.1 Typical antenna configurations.

The interaction of antennas and their shields with the edge plasma is a problem of primary concern in all RF heating experiments. An electrostatic shield is normally required to avoid coupling to electrostatic waves and to avoid contamination of the plasma by other antenna components. An unshielded, all metal antenna will act as a Langmuire probe, drawing current directly through the plasma along field lines passing through the antenna. Even if the antenna is shielded, it can still draw a small current from the tenuous plasma existing in the space between the antenna and the shield. If the antenna is allowed to float, to isolate it from the vacuum vessel, it will charge to a negative DC potential, proportional to the RF voltage. The floating potential, which may be several kV in high power experiments is such that the saturation ion current (drawn during negative half cycles) just balances the electron current (drawn during positive half cycles).

17.2 Matching networks

At wave frequencies in the range 1 – 30 MHz, suitable matching networks for antennas can be constructed from lumped LC networks. At higher frequencies and at high power levels (> 500 kW), transmission line tuning stubs are more commonly used. Matching networks are needed in CW applications to couple maximum power from the generator to the antenna and also to avoid high voltage breakdown problems which can arise from reflections at a badly matched load. Most CW sources are designed to feed a 50Ω load via 50Ω characteristic impedance coaxial cable. The forward and reflected power is generally monitored with a VSWR meter.

An antenna loop used to excite Alfven waves typically has an inductance less than 2 μH and a resistance less than 2Ω. Two circuits commonly used to convert this load into a 50Ω purely resistive load are shown in Fig. 17.2.

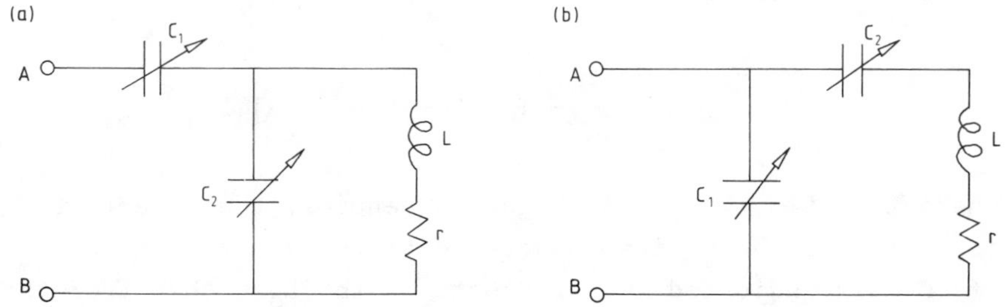

Fig. 17.2 Antenna matching networks.

Consider first the circuit with C_2 in parallel with the antenna (Fig. 17.2a). The impedance of the parallel section is given by

$$\frac{1}{Z} = \frac{1}{j\omega L + r} + j\omega C_2 = \frac{1 - \omega^2 LC_2 + j\omega C_2 r}{j\omega L + r}$$

so $$Z = \frac{r + j\omega L(1 - \omega^2 LC_2) - j\omega C_2 r^2}{(1 - \omega^2 LC_2)^2 + (\omega C_2 r)^2}$$

The imaginary part of Z is zero when

$$\omega^2 = \omega_0^2 = \frac{1}{LC_2} - \frac{r^2}{L^2}$$

and then $Z = L/(rC_2)$. In general, $Z = R + j\omega L_1$ where R and L_1 vary with frequency. Typically $R \gg 50\Omega$ at the resonance frequency, ω_0, and R decreases to $R = r$ at $\omega = 0$. If the frequency (or equivalently C_2) is adjusted to a frequency slighly less than ω_0, then R can be set to 50Ω. C_1 is adjusted to "tune out" the inductive part of Z. The impedance, Z_{AB}, between A and B is

$$Z_{AB} = \frac{1}{j\omega C_1} + R + j\omega L_1 = 50\Omega$$

when $R = 50\Omega$ and $\omega L_1 = 1/\omega C_1$.

In the circuit shown in Fig. 17.2b, the impedance of the series C_2Lr arm is

$$Z = \frac{1}{j\omega C_2} + j\omega L + r = r + j\omega L_0$$

where $L_0 = L - 1/\omega C_2$. At high frequencies, the effect of C_2 is to reduce the series inductance of the arm from L to L_0. If C_1 is adjusted to resonate with L_0, then from the previous results, $Z_{AB} = L_0/(rC_1)$. At resonance, Z_{AB} is purely resistive and can be set to 50Ω by appropriate choice of C_1 and C_2. If Z_{AB} is set to 50Ω under vacuum conditions, it will change to a different value when the antenna is loaded by the plasma. Sometimes it is useful to set Z_{AB} so that it changes to 50Ω during the discharge.

17.3 Transmission lines

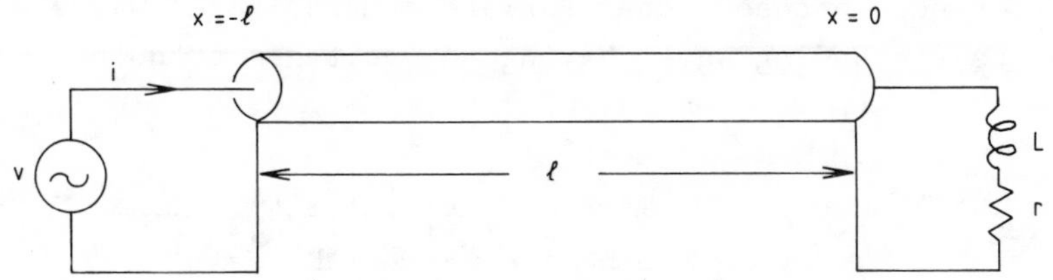

Fig. 17.3 Transmission line of length ℓ.

Suppose that the antenna is connected to a generator or to a matching network via a length, ℓ, of coaxial cable. The characteristic impedance of the cable is Z_o and the phase velocity along the cable is $v_p = \omega/k = c/\sqrt{K}$ where K is the dielectric constant of the medium separating the inner and outer conductors. The voltage and current along the cable are related by

$$v = v_F\, e^{j(\omega t-kx)} = Z_o i \qquad \text{for a forward wave}$$

$$v = v_R\, e^{j(\omega t+kx)} = -Z_o i \qquad \text{for a reflected wave}$$

In general we must sum both forward and reflected waves to give

$$v = v_F\, e^{j(\omega t-kx)} + v_R\, e^{j(\omega t+kx)}$$

and $$Z_o i = v_F\, e^{j(\omega t-kx)} - v_R\, e^{j(\omega t+kx)}$$

For an antenna of impedance Z_L at x = 0,

$$\left(\frac{v}{Z_o i}\right)_{x=0} = \frac{Z_L}{Z_o} = \frac{v_F + v_R}{v_F - v_R}$$

so $$\frac{v_R}{v_F} = \frac{Z_L - Z_o}{Z_L + Z_o}$$

There is no reflected wave, and $v_R = 0$, when $Z_L = Z_0$. If the cable is not terminated in its characteristic impedance, interference between the forward and reflected waves will generate a standing wave having a voltage standing wave ratio

$$\text{VSWR} = \frac{v_{max}}{v_{min}} = \frac{|v_F| + |v_R|}{|v_F| - |v_R|} = \frac{1 + \rho}{1 - \rho}$$

where $\rho = |v_R|/|v_F|$ is the reflection coefficient.

$$\text{The forward power} \quad p_F = |v_F|^2/(2Z_0)$$

$$\text{The reflected power} \quad p_R = |v_R|^2/(2Z_0)$$

The difference between the forward and reflected power is the power transferred to the antenna. The fraction, f, of the forward power dissipated in the antenna is

$$f = \frac{p_F - p_R}{p_F} = 1 - \rho^2 .$$

Normally, one tries to keep VSWR < 3. At VSWR = 3, $\rho = 0.5$ and f = 0.75. If the cable is open circuited at the far end, $Z_L = \infty$, $\rho = 1.0$, VSWR = ∞ and f = 0. When $Z_L = \infty$, the wave is reflected with $v_R = v_F$ and the voltage at the open end is $v_F + v_R = 2v_F$.

At the input end of the cable, $x = -\ell$,

$$\frac{Z_{in}}{Z_0} = \left(\frac{v}{Z_0 i}\right)_{x=-\ell} = \frac{v_F\, e^{jk\ell} + v_R\, e^{-jk\ell}}{v_F\, e^{jk\ell} - v_R\, e^{-jk\ell}}$$

where Z_{in} is the input impedance.

$$\therefore \quad \frac{Z_{in}}{Z_0} = \frac{Z_L + jZ_0 \tan(k\ell)}{Z_0 + jZ_L \tan(k\ell)} = r_{in} + jx_{in}$$

where $\quad r_{in} = r_L(1 + \tan^2\theta)/D$

$$x_{in} = [x_L(1-\tan^2\theta) + (1-x_L{}^2-r_L{}^2)\tan\theta]/D$$

$$D = (1-x_L\tan\theta)^2 + r_L{}^2\tan^2\theta$$

$$Z_L = R_L + jX_L \qquad r_L = R_L/Z_o \qquad x_L = X_L/Z_o$$

$$\theta = k\ell = 2.09 \times 10^{-8}\, \ell\, f\, \sqrt{K} \quad \text{(radians)}$$

ℓ in metres, f in Hz and $\sqrt{K} \sim 1.5$ for most 50Ω cables. These results show that the input impedance is the same as Z_L at low frequencies, but increases towards infinity when the cable is approximately one quarter wavelength long. For example, if $\ell = 2$ metre, $Z_{in} \to \infty$ at f ~ 8 MHz (depending on R_L and X_L).

One of the most important measurements for wave heating experiments is the effective series resistance, R_L, of the antenna, since this determines the power dissipated in the plasma. If p_F and p_R are measured with directional couplers (eg a VSWR meter) and the current i_L in the antenna is measured with a current transformer, then

$$R_L = 2(p_F - p_R)/|i_L|^2 .$$

In order to measure both the antenna resistance and inductance, it is better to measure the antenna voltage and current directly to obtain the necessary amplitude and phase information. The inductance can change significantly, especially in the presence of cavity modes, due to the change in magnetic flux linking the antenna. The effective inductance can even be negative if a large reverse flux links the antenna.

17.4 Current transformers

Current transformers for use at RF frequencies are readily available from commercial sources, but home-made versions are cheaper and easy to construct (Fig. 17.4).

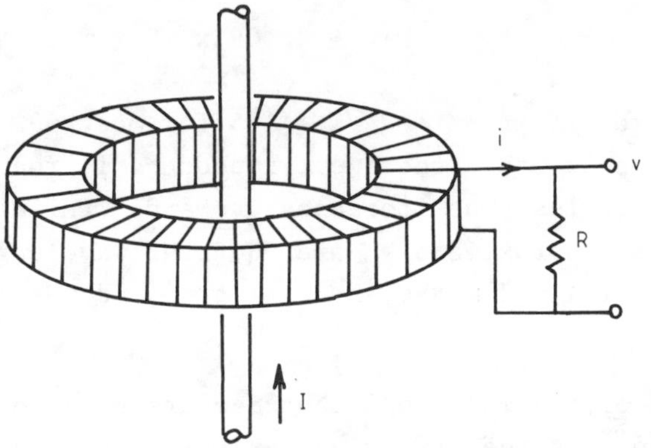

Fig. 17.4 Current transformer used to measure I.

A current transformer for use at high frequencies consists of a coil of N turns wound on a toroidal ferrite core and connected to a load resistance $R \le 50\Omega$. Since $i = I/N$, $v = RI/N$. This voltage signal is proportional to I and in phase with I. For example, if $I = 100$ A, $N = 100$, $R = 1\Omega$, then $v = 1$ volt. For good high frequency response, the core and number of turns should be small, consistent with possible voltage breakdown to the primary lead. If possible, the transformer should therefore be located in a low voltage section of the circuit. This also reduces electrostatic pick-up problems due to the capacitance between the primary and secondary.

With large cores, the high frequency response can be improved considerably by distributing the load circumferentially around the coil to avoid standing wave resonances. Due to the distributed inductance and capacitance of the winding, the secondary coil behaves as a transmission line, and supports standing waves due to reflections at each end of the line. The resonances are strongly damped if R is made from say ten series resistors, each resistor being soldered across N/10 turns, as shown in Fig. 17.5. Other design criteria are given in Refs. L1 and L4.

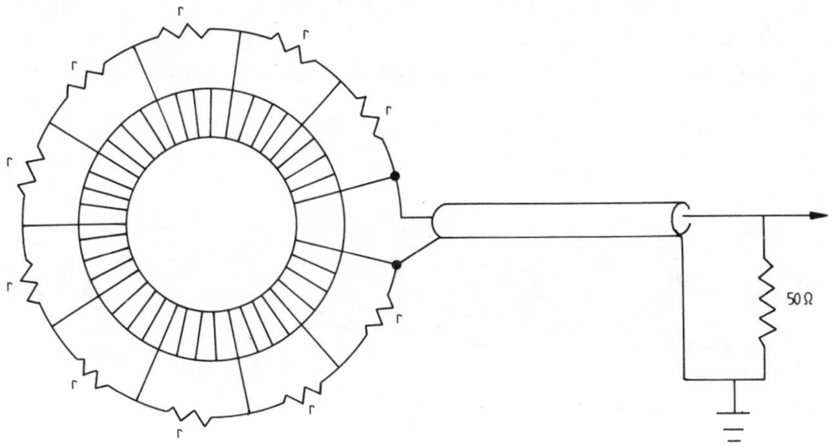

Fig. 17.5 A current transformer with a distributed load.

In plasma physics experiments, it is important to avoid saturation of the transformer core by keeping it away from strong magnetic fields. Alternatively, an air-core coil can be used. It will act as a current transformer if $\omega L \gg R$ or as a Rogowski coil if $\omega L \ll R$, where L is the coil inductance. The output voltage of a Rogowski coil is proportional to dI/dt and therefore needs to be integrated to obtain a signal proportional to I. Current transformers are self integrating in the sense that the induced current i is in phase with I, provided $\omega L \gg R$. If $R \gg \omega L$, the induced current is given by $i = (NAdB/dt)/R$ and hence i and I are 90° out of phase. The general expression for i, valid for arbitrary R or L, is

$$i = \frac{NAdB/dt}{R + j\omega L}$$

17.5 Magnetic probes

A magnetic probe consists of a small coil of area A, with N = 1–30 turns of fine wire wound onto a small insulating former and fed to a detector or oscilloscope via coaxial cable or twisted leads. A probe array containing more than four coils usually requires twisted lead outputs due to space limitations. The leads must be twisted tightly to avoid any loops other than in the measuring coil, and the final product tested to see if there is any pickup along the probe stem.

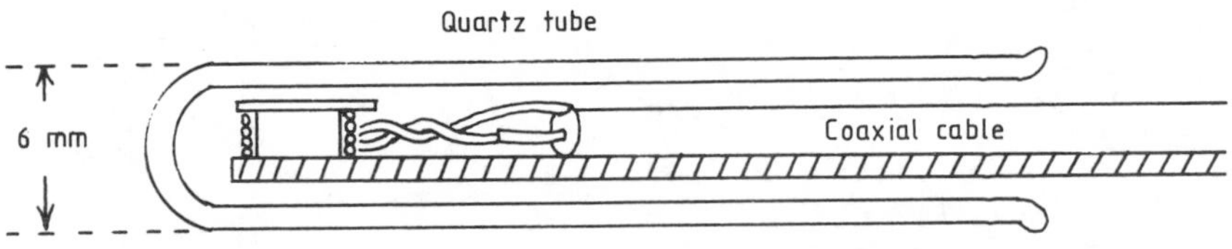

Fig. 17.6 Typical magnetic probe construction.

Since the output voltage from the probe coil is $v = NAdB/dt$, the sensitivity of the probe increases with the coil area and number of turns. The induced emf drives a current $i = v/R$ in the load, R, in a direction which acts to weaken the B field inside the coil. Consequently, the effective value of NA decreases at high frequencies, with a −3dB frequency at $\omega = R/L$ where L is the coil inductance. If $L = 1\ \mu H$, $R = 50\Omega$, the frequency response is flat up to about 4 MHz. The coil could still be used at say 10 MHz, but a significant phase shift would be introduced, on top of the phase delay caused by the transmission time of the signal along the coaxial cable. Due to the latter effect, it is desirable to ensure that cable lengths are matched when comparing the outputs of two or more probe coils.

It is usually not necessary to shield a probe coil or twisted leads from plasma noise, since the noise is generally at a much lower frequency and can be filtered out. However, the coil and leads are then sensitive to electric components of the wave fields which couple capacitively to the probe. The latter effect can be tested by rotating the probe through 180° to see if the output signal reverses phase. If not, there are three solutions; (a) shield the coil or leads, (b) use two counter-wound or centre-tapped coils in close proximity and subtract the outputs differentially, or (c) use a hybrid combiner to separate the magnetic and electric field components from a single, floating coil. A simple combiner circuit is shown in Fig. 17.7 (G2). This circuit uses three identical coils, each with about 20 turns, trifilar wound on a small toroidal ferrite core (ie three separate wires are first twisted together then wound like a single coil onto the core). Such a combiner is easily constructed and the frequency response is flat to about 30 MHz.

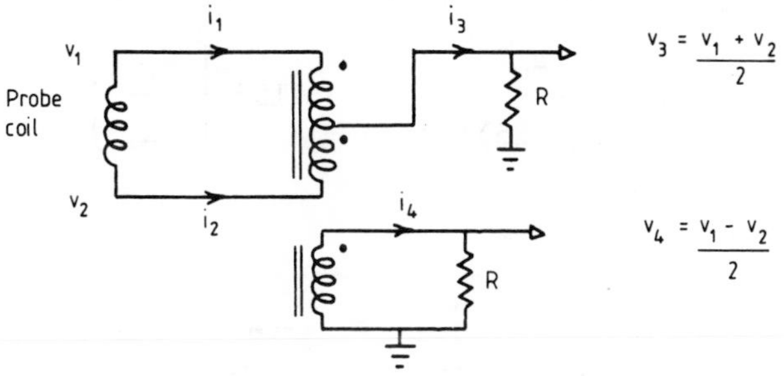

Fig. 17.7 Hybrid combiner used to separate the electric and magnetic components from a single probe coil. The input impedances are $v_1/i_1 = v_2/i_2 = 2R$. The magnetic signal is v_4, while v_3 gives the electric component.

The voltage relationships are derived as follows. Let v be the voltage drop across each of the three transformer windings. Then

$$v_1 = v + v_3 = v_4 + v_3$$

$$v_2 = -v + v_3 = -v_4 + v_3$$

$$\therefore \quad v_3 = (v_1 + v_2)/2 \quad \text{and} \quad v_4 = (v_1 - v_2)/2$$

The values of i_1 and i_2 can be calculated on the assumption that the total flux in the core remains small. This is the assumption used to derive the standard relation $N_1 i_1 = N_2 i_2$ for the currents in the primary and secondary windings of a transformer. For the hybrid combiner, $i_1 - i_2 - i_4 = 0$ where $i_1 + i_2 = i_3$, $Ri_3 = v_3$ and $Ri_4 = v_4$. Hence $i_1 = v_1/2R$ and $i_2 = v_2/2R$. These relations are valid provided $\omega L \gg R$ where L is the inductance of each winding. The condition $\omega L = R$ gives the low frequency −3dB point. Typically, R = 25Ω or 50Ω.

17.6 Probe calibration

The sensitivity and frequency response of a magnetic probe (usually terminated with 50Ω) can be determined using a known, uniform magnetic field. A suitable source is a Helmholtz coil, shown in Fig. 17.8. The field is very uniform in the central region and is given by

$$B = \frac{\mu_0 I}{a(1.25)^{1.5}} = 0.7155 \left(\frac{\mu_0 I}{a} \right)$$

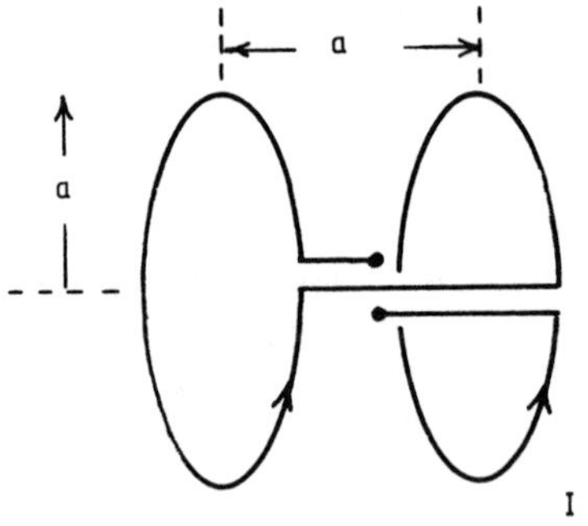

Fig. 17.8 Helmholtz coil

At high frequencies, the Helmholtz coil will resonate due to its self - capacitance. Consequently, one has to be careful that the current measured external to the coil is the same as the current in the coil. If the coil has only one turn in each arm and if the input leads are not twisted together, then the self-resonance frequency can be made to exceed 200 MHz. This can be tested at lower frequencies by soldering various fixed capacitors (~ 10 pF) in parallel with the coil and extrapolating the results to calculate the self resonance frequency without any added capacitance.

An array of coils can be calibrated by inserting the array in a long solenoid or in a thin, rectangular cross-section solenoid, depending on the coil orientation. It is advisable to rotate the probe through 90° to ensure that the signal drops to zero when the coil axis is perpendicular to the source field, otherwise the results would be meaningless.

17.7 Filters

A simple LC filter designed to eliminate noise from probe signals is shown in Fig. 17.9. The circuit is designed to resonate at the operating frequency and has an input impedance about 50Ω over a wide frequency range. Since L and C are in series, the voltage across C is larger than the input signal, so the filter has a passive gain. The gain depends on the circuit Q. Noise spikes will also set the circuit into oscillation, but the oscillations are quickly damped if the Q is not too high. A diode detector can be connected to the output if desired, to monitor the wave amplitude, without degrading the Q significantly.

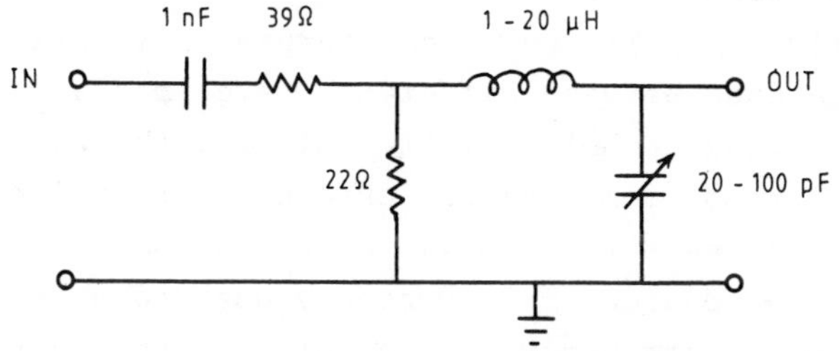

Fig. 17.9 Band pass filter with passive gain.

Another useful filter is the notch filter shown in Fig.17.10. It acts as a band - stop filter and is therefore useful for rejecting unwanted signals (eg sub or second harmonics). The output signal is zero at a frequency

$$\omega_o = \frac{\sqrt{n}}{R_1 C_1} \qquad \text{where} \quad n = \frac{2C_1}{C_2} = \frac{R_1}{2R_2}$$

Usually, n = 1. For example, $f_o = 4.0$ MHz if $C_1 = 100$ pF, $R_1 = 390\Omega$ or if $C_1 = 1$ nF, $R_1 = 39\Omega$. The choice n = 1 is convenient since then $C_2 = 2C_1$ and $R_2 = R_1/2$.

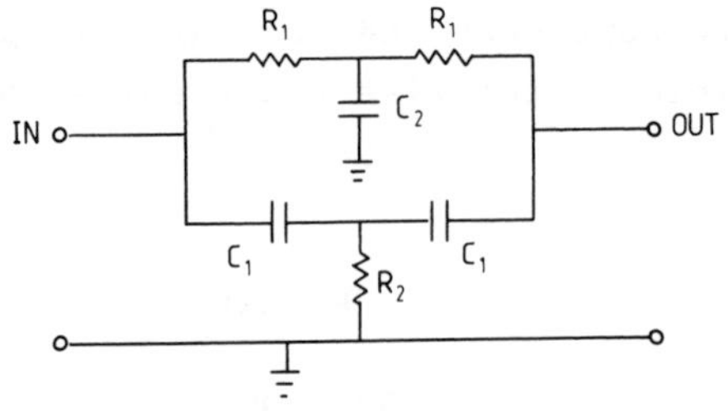

Fig. 17.10 Band-stop or notch filter.

17.8 Phase detectors

A lot of progress can be made in Alfven wave experiments by observing only the amplitude of the wave magnetic field components, together with other information obtained from laser scattering or from measurements of the antenna impedance. The phase of a wave signal (with respect to the antenna current for example) is more difficult to measure and usually more difficult to interpret, especially if there is interference between different wave modes. A suitable phase measuring technique is to multiply the wave signal by a signal proportional to the antenna current (or some other signal of the same frequency) using a balanced mixer. A mixer is a device most commonly used to obtain an output signal whose frequency is equal to the difference in frequency between two input signals. The two input signals are known as the RF and LO (local oscillator) signals and the output is called the IF (intermediate frequency) signal. If the two input signals are of the same frequency, the output signal is a DC signal proportional to $\cos\phi$ where ϕ is the phase difference between the two input signals. Suppose the two input signals are

$$v = v_0 \sin(\omega t + \phi) \quad \text{and} \quad i_1 = i_0 \sin(\omega t)$$

then

$$vi_1 = (v_0 i_0/2)[\cos\phi - \cos(2\omega t + \phi)]$$

The mixer performs this multiplication either electronically or passively depending on its design. The output signal is proportional to $\cos\phi$ but also contains a component at frequency 2ω. The high frequency component must be filtered out to measure $\cos\phi$. If the antenna current signal, i_1, is 90° phase shifted to produce a signal $i_2 = i_0 \cos(\omega t)$, then multiplication of v and i_2 yields

$$vi_2 = (v_0 i_0/2)[\sin\phi + \sin(2\omega t + \phi)]$$

Consequently, two mixers can be used to monitor $\sin\phi$ and $\cos\phi$ simultaneously to determine ϕ uniquely in the range 0 to 360°. To avoid sensitivity to amplitude variations in v or i, these signals should first be amplified to produce constant amplitude square waves which are then multiplied. Such a system can also be used to measure the complex antenna impedance by multiplying the antenna current and voltage waveforms.

Mixers and phase detectors are available commercially. The most common variety is passive, constructed from transformers and a ring of diodes. The output signal depends on $\cos\phi$ but it is not linearly proportional to $\cos\phi$. A linear, electronic version is described by Borg (L3).

17.9 RC Compensated integrator

Integrators are commonly used, in plasma physics experiments, to convert a coil signal NAdB/dt into a signal proportional to B. The simplest type of integrator consists of a resistor, R, and capacitor, C, in series. The voltage across C is given by $v_c = q/C = \int v_{in}\, dt/RC$, provided RC ≫ (time of interest) or provided $\omega RC \gg 1$ for sinusoidal signals. The output voltage is much smaller than the input voltage. If the output signal is less than a few mV, one solution is to use an operational amplifier integrator to obtain a larger signal. An alternative solution is shown in Fig. 17.11. This circuit (L5) gives an output signal about ten times larger than a simple RC integrator (for the same amount of integrator droop) and is also a much better 90 degree phase shifter.

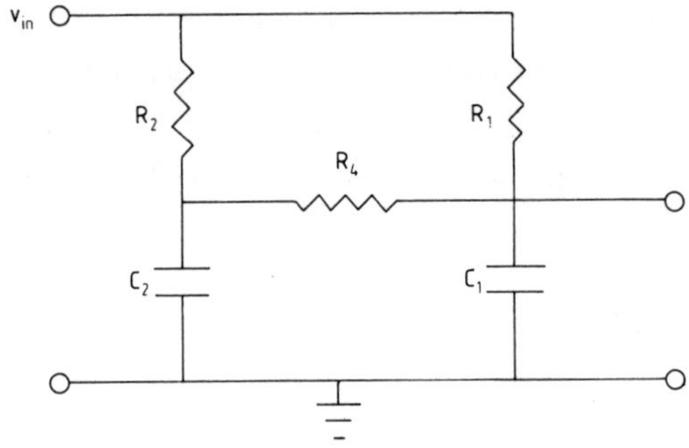

$$R_4 = R_1 = 2R_2$$

$$C_1 = C_2$$

$$v_{out} = \frac{1}{R_1 C_1} \int v_{in}\, dt$$

Fig. 17.11 RC compensated integrator.

SPACE FILLER

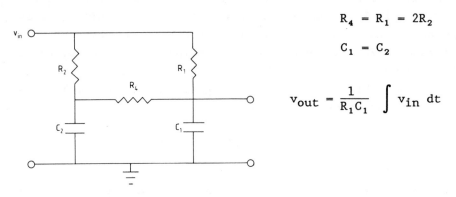

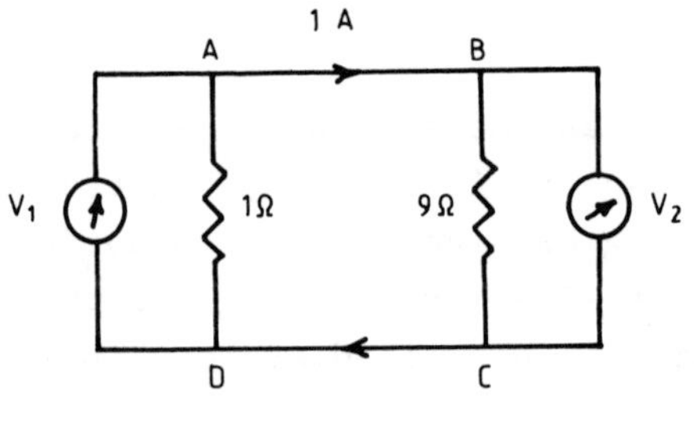

An emf of 10V is induced by a changing flux in the loop ABCD. The total resistance of the loop is 10Ω, so the induced current is 1A. What are the readings on the voltmeters V_1 and V_2 ? Note that the meters are connected to the same points so they might be expected to read the same.

COMMENTS ON SPACE FILLERS

Page 13. See Aström E. (1950) *Arkiv för Fysik*, **2**, 443 and Herlofson N. (1950) *Nature* **165**, 1020.

Page 16. The concepts of magnetic pressure and tension arise from the relation

$$\mathbf{j} \times \mathbf{B} = (\nabla \times \mathbf{B}) \times \mathbf{B}/\mu_0 = (\mathbf{B}\cdot\nabla)\mathbf{B}/\mu_0 - \nabla(B^2/2\mu_0)$$

The first term is due to tension and the second term is the magnetic pressure gradient. Consider the case where $\mathbf{B} = \hat{\theta}\, B_\theta$. Then

$$(\nabla \times \mathbf{B}) \times \mathbf{B}/\mu_0 = -\,\hat{\mathbf{r}} \left[\frac{B_\theta{}^2}{\mu_0 r} + \frac{\partial}{\partial r}\left(\frac{B_\theta{}^2}{2\mu_0}\right) \right]$$

Even though the tension is directed along the field lines, the effect of the tension is to produce a *radial* force inversely proportional to the radius of curvature of the field lines. It acts like the tension in a rubber band. Elementary texts on waves give an analogous expression for the restoring force in a stretched string.

At any point in a vacuum, where $\mathbf{j} = 0$, the magnetic force, $\mathbf{j} \times \mathbf{B}$, is zero. The tension and pressure terms are then equal and opposite. A current–carrying wire placed in a uniform magnetic field experiences a force due to the forces acting *inside* the wire. In fact, if the current density is uniform, the tension and pressure forces are exactly equal in magnitude and direction at all points inside the wire [see Cross R.C. (1989) *Am. J. Phys.* in press].

Page 28. See Zelby L.W. (1967) *Am. J. Phys.* **35**, 1094.

Page 36. Since $\nabla\cdot\mathbf{E} = (n_i - n_e)/\epsilon_0$, there will be a small difference between n_i and n_e for the ideal MHD torsional wave. One can estimate the difference from the single fluid equations, using $n_i - n_e = ik_xE_x\epsilon_o/e$. When $b/B \sim 0.01$ and $k_x \sim 500\ m^{-1}$, then $n_i - n_e \sim 10^{-4}\ n_e$.

Page 46. The water wave picture *is* a good representation of an Alfven surface wave. The wave fields decrease exponentially away from the surface and the fluid velocity, from Eqs.(3.4) and (3.11) is given by $v_x/v_y = b_x/b_y = k_x/k_y$. If $k_y \gg k_z$, then $k_x = \pm ik_y$ and $v_x = \pm iv_y$. At any given point, v_x and v_y are equal in amplitude and 90° out of phase. At any given instant of time, the velocity field is elliptical, as drawn [see Main I.G. (1978) *Vibrations and waves in physics*, Cambridge Univ. Press, Chapter 13].

Page 58. (a) b_r *does* change sign at $r = 0$, as shown in Fig. 9.1. (b) An antenna phased to generate an $m = +1$ wave will generate an $m = +1$ wave in both the +z and the −z directions. If the direction of **B** is reversed, the same antenna will launch $m = -1$ waves in both directions. However the coupling efficiency is different for $m = +1$ and $m = -1$ waves. Antennas can also be designed to couple preferentially to +z or −z propagation (C7, F1).

Page 70. When $E_z = 0$, the phase velocity of a plane wave in the z direction is given, from Eq. (3.6), by $\omega/k_z = E_x/b_y$. This ratio can be regarded as a form of impedance, in the sense that E is proportional to voltage and b is proportional to current. If a given E drives a large current, then ω/k is small. In free space, E drives a small displacement current, so $\omega/k = c$ is large. In an Alfven wave, E drives a large conduction current and $\omega/k_z = v_A$ is small.

In ideal MHD, an E_x field generates a v_y drift in the presence of a B_z field. If E_x is a DC field, there is no j_x current. In an alternating field, a j_x current will be produced by relative displacement of ions and electrons in the x direction. The transverse j_x current in an Alfven wave is carried mainly by the ions, due to their large gyroradius. The ions are accelerated by the torsional wave, especially as $\Omega_i \to 1$, and therefore act to reduce the wave impedance as $\Omega_i \to 1$. For a plane wave, with $k_\perp = 0$, $E_y = -iE_x$ for the torsional wave and $i\mu_0 j_x = \omega E_x/v_A{}^2(1 - \Omega_i)$. As $\Omega_i \to 1$, $j_x/E_x \to \infty$, and the phase velocity drops to zero.

Page 128. (a) The complex number $E_R e^{-i\omega t}$ can be represented by a vector of length E_R rotating clockwise at angular frequency ω, with components $E_x = E_R\cos(\omega t)$ and $E_y = -E_R\sin(\omega t)$. The complex number $E_L\exp(+i\omega t)$ rotates anticlockwise, with components $E_x = E_L\cos(\omega t)$ and $E_y = E_L\sin(\omega t)$. The sum of the two fields has an x component $(E_L + E_R)\cos(\omega t)$ and a y component $(E_L - E_R)\sin(\omega t)$. The resultant field is elliptically polarised. It can be represented by a vector which changes its length from $E_L - E_R$ to $E_L + E_R$ as it rotates. The field rotates clockwise if $E_R > E_L$ or anticlockwise if $E_L > E_R$.

(b) If a wave of amplitude A is launched by an antenna in a torus, then after one toroidal trip, it arrives with amplitude $A\eta$ and phase shift ϕ. The resulting signal at this time is $A + A\eta\exp(i\phi)$. After an infinite time, the steady state signal is

$$y = A + A\eta\, e^{i\phi} + A\eta^2\, e^{2i\phi} + \ldots \quad = \quad A/[1 - \eta\exp(i\phi)]$$

since $1/(1 - x) = 1 + x + x^2 + x^3 + \ldots$

Now consider the fact that waves travel in both directions toroidally. Assume that the wave amplitude varies with distance, z, as $y = A\exp(-z/\ell)$ where ℓ is the attenuation length. At $z = 2\pi r$, $\exp(-2\pi r/\ell) = \eta$. Consider also a point displaced toroidally by an angle ϕ_o from the antenna, or at distances $r\phi_o$ and $r(2\pi - \phi_o)$ in each direction from the antenna. The steady state signal at this point is given by

$$y = A(B + C)(1 + \eta\, e^{i\phi} + \eta^2\, e^{2i\phi} + \ldots)$$

where $B = e^{-r\phi_o/\ell}\, e^{i\phi_o\phi/2\pi}$

and $C = e^{-r(2\pi - \phi_o)/\ell}\, e^{i(2\pi - \phi_o)\phi/2\pi}$

where ϕ is the phase shift of the wave in a distance $z = 2\pi r$.

Page 160. If $\mathbf{B} = \hat{\theta}\, B_\theta$ then the ideal MHD equations give $E_\theta = 0$. Let **E** and **b** vary as $f(r)\exp[i(k_z z + m\theta - \omega t)]$. The dispersion relation for the torsional wave is obtained by setting $b_\theta = 0$, in which case $\omega/k_\theta = v_A$ where $k_\theta = m/r$. The

Poynting vector $\mathbf{E} \times \mathbf{b}/\mu_o = \hat{\theta}(E_z b_r - E_r b_z)$ is parallel to **B**. See also Kuo S.P. et. al. (1987) *J. Plas. Phys.* **38**, 235 and Swanson D.G. (1975) *Phys. Fluids* **18**, 1269 and **17**, 2241.

Page 175. (i) It is fairly obvious that a small b_r field cannot alter the strength of the toroidal field. The field strength can be represented by the number of field lines passing through unit area only if the field lines are all different field lines. In Fig. 15.6b, the same field line passes through the plasma cross section many times. The spacing between the dots in the Figure does not represent the field strength. It simply represents the displacement of a field line after each toroidal trip.

(ii) It appears likely that electrons will be guided along field lines towards the X-points, but charge separation will prevent an infinite accumulation of charge.

(iii) The island structure shown in Fig. 15.6b is formed by a $k_{\parallel} = 0$ current perturbation. In other words, the current is essentially a DC current flowing along the field lines. The Alfven wave, on the other hand, is a finite $k_{\parallel}$ perturbation. In ideal MHD, an Alfven resonance is defined by the relation $\omega = k_{\parallel} v_A = (n + m/q)v_A/R$. Consequently, $k_{\parallel}$ is finite when ω is real and finite, so b_r reverses sign periodically along any given steady magnetic field line. As a result, the radial displacement of a field line, after each toroidal trip, will be much smaller than that produced by a $k_{\parallel} = 0$ perturbation of the same amplitude.

Page 198. V_1 reads 1 volt and V_2 reads 9 volts, despite the fact that both meters are connected to the same points. Potential difference is not a uniquely defined quantity for an induced electric field, since it depends on how the measurement is made. For example, V_1 is connected across the 1Ω resistor and measures 1V. V_2 is also connected across the 1Ω resistor but there is an additional emf of 10V in the loop formed by the leads connecting V_2 to the 1Ω resistor. Hence, V_2 reads 1V − 10V = −9V. It is easy to set up the circuit using a toroidal transformer core.

REFERENCES

A. Books, conference proceedings

B. Early theoretical work

C. Early experimental work

D. Alfven surface waves

E. The Alfven wave heating scheme

F. Alfven wave heating experiments

G. Magnetically guided Alfven waves

H. Alfven wave spectrum

I. ICRF heating (theory)

J. ICRF heating experiments

K. Magnetoacoustic waves

L. Experimental techniques

M. Space plasmas

A. Books and Conference Proceedings

(a) Books

1. Booker H.G. (1984) *Cold plasma waves*, Martinus Nijhoff, Dordrecht.

2. Boyd T.J.M. and Sanderson J.J. (1969) *Plasma Dynamics*, Thomas Nelson, London.

3. Budden K.G. (1961) *Radio waves in the ionosphere*, University Press, Cambridge.

4. Galejs J. (1969) *Antennas in inhomogeneous media*, Pergamon, Oxford.

5. Granatstein V.L. and Colestock P.L. (Eds) (1985) *Wave heating and current drive in plasmas*, Gordon and Breach, New York.

6. Hasegawa A. and Uberoi C. (1982) *The Alfven wave* , Technical Information Center, U.S. Dept. of Energy.

7. Spitzer L. (1962) *Physics of Fully Ionized Gases*, Interscience, New York.

8. Stix T.H. (1962) *The theory of Plasma Waves*, McGraw-Hill, New York.

(b) Conference Proceedings

9. Proc. 4th Int. Symp. on heating in toroidal plasmas, Rome, 1984.

10. Proc. 3rd Int. Symp. on heating in toroidal plasmas, Grenoble, 1982.

11. 14th European Conf. on Contr. Fusion and Plasma Heating, Madrid, 1987.

12. 13th European Conf. on Contr. Fusion and Plasma Heating, Schliersee, 1986.

13. Proc. 10th Int. Conf. on Plasma Phys. and Contr. Nucl. Fus. Research, London (1984) IAEA, Vienna, 1985.

14. 6th Topical Conf. on Radiofrequency Plasma Heating, Callaway Gardens (1985), AIP Conf. Proc. Number 129, Am. Inst. Phys. New York.

B. Early theoretical work

1. Alfven H. (1942) Existence of Electromagnetic-Hydrodynamic Waves, *Nature* **150**, 405.

2. Astrom E. (1950) Magnetohydrodynamic waves in a plasma, *Nature* **165**, 1019.

3. Astrom E. (1950) On waves in an ionised gas, *Ark. Fys.* **2**, 443.

4. Bernstein I.B. (1958) Waves in a plasma in a magnetic field, *Phys. Rev.* **109**, 10.

5. Buchsbaum S.J. (1960) Resonance in a plasma with two ion species, *Phys. Fluids* **3**, 418.

6. Ferraro V.C. (1954) On the reflection and refraction of Alfven waves, *Astrophys. J.* **119**, 393.

7. Formisano V. and Kennel C.F. (1969) Small amplitude waves in high β plasmas, *J. Plasma Phys.* **3**, 55.

8. Lighthill M.J. (1960) Studies on magnetohydrodynamic waves and other anisotropic wave motions, *Phil. Trans. Roy. Soc.* **A252**, 397.

9. Newcomb W.A. (1957) The hydromagnetic wave guide, in *Magnetohydrodynamics* , R. Landschoff (Ed.), Stanford University Press.

10. Pneuman G.W. (1965) Hydromagnetic waves in a nonuniform current carrying plasma column, *Phys. Fluids* **8**, 507.

11. Roberts P.H. (1955) On the reflection and refraction of hydro-dynamic waves, *Astrophys. J.* **121**, 720.

12. Stein R.F. (1970) Reflection, Refraction and coupling of MHD waves at a density step, *Astrophys. J. Supp. Series* **192**, 419.

13. Stix T.H. (1965) Radiation and absorption via mode conversion in an inhomogeneous collision-free plasma, *Phys. Rev. Lett.* **15**, 878.

14. Stringer T.E. (1963) Low frequency waves in an unbounded plasma, *Plasma Phys. (Jnl. Nuclear Energy, Part C)* **5**, 89.

15. Woods L.C. (1962) Hydromagnetic waves in a cylindrical plasma, *J. Fluid Mech.* **13**, 570.

C. Early experimental work

1. Allen T.K., Baker W.R., Pyle R.V. and Wilcox J.M. (1959) Experimental generation of plasma Alfven waves, *Phys. Rev. Lett.* **2**, 383.

2. Cross R.C. and Lehane J.A. (1968) Hydromagnetic waves in a nonuniform hydrogen plasma, *Phys. Fluids* **11**, 2621.

3. Hooke W.M. and Rothman M.A. (1964) Review. A survey of experiments on ion cyclotron resonance in plasmas, *Nuclear Fusion* **4**, 33.

4. Hosea J.C. and Sinclair R.M. (1970) Ion cyclotron wave generation in the Model C stellarator, *Phys. Fluids* **13**, 701.

5. Jephcott D.F. (1959) Alfven waves in a gas discharge, *Nature* **183**, 1652.

6. Jephcott D.F. and Stocker P.M. (1962) Hydromagnetic waves in a cylindrical plasma, *J. Fluid Mech.* **13**, 587.

7. Knox S.O., Paoloni F.J. and Kristiansen M. (1975) Helical antenna for exciting azimuthally asymmetric Alfven waves, *J. Appl. Phys.* **46**, 2516.

8. Kristiansen M. and Dougal A.A. (1969) Stop- and pass-bands for harmonic ion cyclotron waves, *Plasma Physics* **11**, 19.

9. Lehnert B. (1954) Magnetohydrodynamic waves in liquid sodium, *Phys. Rev.* **94**, 815.

10. Lundquist S. (1949) (a). Experimental demonstration of magnetohydrodynamic waves, *Nature* **164**, 145. (b). Experimental investigations of magnetohydrodynamic waves, *Phys. Rev.* **76**, 1805.

11. McPherson D.A. and Pridmore-Brown D.C. (1966) Density gradient effects on Alfven wave propagation in a cylindrical plasma, *Phys. Fluids* **9**, 2033.

12. Muller G. (1974) Experimental study of torsional Alfven waves in a cylindrical partially ionized magnetoplasma, *Plasma Physics* **16**, 813.

13. Paoloni F.J. (1973) Non-axisymmetric Alfven wave propagation in an argon plasma, *Plasma Phys.* **15**, 475.

14. Wilcox J.M., Boley F.I. and DeSilva A.W. (1960) Experimental study of Alfven wave properties, *Phys. Fluids* **3**, 15.

15. Wilcox J.M., DeSilva A.W. and Cooper III. W.S. (1961) Experiments on Alfven wave propagation, *Phys. Fluids* **4**, 1506.

D. Alfven surface waves

1. Appert K., Vaclavik J. and Villard L. (1984) Spectrum of low-frequency, non axisymmetric oscillations in a cold, current-carrying plasma column, *Phys. Fluids* **27**, 432.

2. Collins G.A., Cramer N.F. and Donnelly I.J. (1984) MHD surface waves in a resistive cylindrical plasma at low to ion cyclotron frequency, *Plasma Phys. Contr. Fusion* **26**, 273.

3. Cramer N.F. and Donnelly I.J. (1984) Surface and discrete Alfven waves in a current carrying plasma, *Plasma Phys. Contr. Fusion* **26**, 1285.

4. Cramer N.F. and Yung C.M. (1986) Surface waves in a two-ion- species magnetized plasma, *Plasma Phys. Contr. Fusion* **28**, 1043.

5. Cross R.C. and Murphy A.B. (1986) Alfven wave modes in a cylindrical plasma with finite edge density, *Plasma Phys. Contr. Fusion* **28**, 597.

6. Donnelly I.J. and Cramer N.F. (1984) MHD surface waves in a cylindrical plasma with a spatial density variation, *Plasma Phys. Contr. Fusion* **26**, 769.

7. Eways S.E. and Oakes M.E. (1985) Low-frequency plasma-waveguide cutoff, *Phys. Fluids* **28**, 444.

8. Lee M.A. and Roberts B. (1986) On the behaviour of hydromagnetic surface waves, *Astrophys. J.* **301**, 430.

9. Paoloni F.J. (1975) Boundary effects on $m = 0, \pm 1$ Alfven waves in a cylindrical collisionless plasma, *Phys. Fluids* **18**, 640.

10. Wentzel D.G. (1979) Hydromagnetic surface waves, *Astrophys. J.* **227**, 319.

11. Wentzel D.G. (1979) The dissipation of hydromagnetic surface waves, *Astrophys. J.* **233**, 756.

E. The Alfven wave heating scheme (theory)

1. Appert K., Gruber R., Troyon F. and Vaclavik J. (1982) Excitation of global eigenmodes of the Alfven wave in tokamaks, *Plasma Phys.* **24**, 1147.

2. Appert K. et al. (1982) MHD computations for Alfven wave heating in tokamaks, *Nuclear Fusion* **22**, 903.

3. Appert K. and Vaclavik J. (1983) Effects of finite ion cyclotron frequency on the absorption of magnetohydrodynamic waves at the spatial Alfven resonance, *Plasma Phys.* **25**, 551.

4. Bernstein I.B. (1983) Alfven wave heating in general toroidal plasmas, *Phys. Fluids* **26**, 730.

5. Chen L. and Hasegawa A. (1974) Plasma heating by spatial resonance of Alfven wave, *Phys. Fluids* **17**, 1399.

6. Connor J.W., Tang W.M. and Taylor J.B. (1983) Influence of gyroradius and dissipation on the Alfven wave continuum, *Phys. Fluids* **26**, 158.

7. Davis C.L. and Donnelly I.J. (1986) Discrete Alfven waves in planar flux tubes, *Astrophys. and Space Science* **127**, 265.

8. Donnelly I.J. and Clancy B.E. (1983) The spatial Alfven resonance in a collisional plasma, *Aust. J. Phys.* **36**, 305.

9. Donnelly I.J., Clancy B.E. and Cramer N.F. (1985) Alfven wave heating of a cylindrical plasma using axisymmetric waves. Part 1. MHD theory, *J. Plasma Phys.* **34**, 227. Part 2. Kinetic theory, *J. Plasma Phys.* **35**, 75.

10. Hasegawa A. and Chen L. (1974) Plasma heating by Alfven wave phase mixing, *Phys. Rev. Lett.* **32**, 454.

11. Hasegawa A. and Chen L. (1975) Kinetic processes of plasma heating due to Alfven wave exitation, *Phys. Rev. Lett.* **35**, 370.

12. Hasegawa A. and Chen L. (1976) Kinetic processes in plasma heating by resonant mode conversion of Alfven wave, *Phys. Fluids* **19**, 1924.

13. Itoh K. and Itoh S.I. (1983) Energy deposition profile of shear Alfven wave heating, *Plasma Phys.* **25**, 1037.

14. Kappraff J.M. and Tataronis J.A. (1977) Resistive effects on Alfven wave heating, *J. Plasma Phys.* **18**, 209.

15. Karney C.F.F., Perkins F.W. and Sun Y.C. (1979) Alfven resonance effects on magnetosonic modes in large tokamaks, *Phys. Rev. Lett.* **42**, 1621.

16. Mahajan S.M., Ross D.W. and Chen G.L. (1983) Discrete Alfven eigenmode spectrum in magnetohydrodynamics, *Phys. Fluids* **26**, 2195.

17. Mahajan S.M. (1984) Kinetic theory of shear Alfven waves, *Phys. Fluids* **27**, 2238.

18. Ott E., Wersinger J.M. and Bonoli P.T. (1978) Theory of plasma heating by magnetosonic mode absorption, *Phys. Fluids* **21**, 2306.

19. Puri S. (1987) Antenna optimisation for Alfven wave heating, *Nuclear Fusion* **27**, 229.

20. Pritchett P.L. and Dawson J.M. (1978) Phase mixing in the continuous spectrum of Alfven waves, *Phys. Fluids* **21**, 516.

21. Ross D.W., Chen G.L. and Mahajan S.M. (1982) Kinetic description of Alfven wave heating, *Phys. Fluids* **25**, 652.

22. Ross D.W., Li Y. and Mahajan S.M. (1986) Antenna analysis and boundary conditions for Alfven wave studies in tokamaks, *Nucl. Fusion* **26**, 139.

F. Alfven wave heating experiments

1. Amagishi Y., Inutake M., Akitsu T. and Tsushima A. (1981) Non-axisymmetric Alfven wave excited by a helical coupler in an inhomogeneous plasma, *Japanese J. Applied Physics* **20**, 2171.

2. Amagishi Y. and Tsushima A. (1984) Formation of spatial Alfven resonance and cutoff fields in an inhomogeneous plasma, *Plasma Phys. Contr. Fusion* **26**, 1489.

3. Appert K. et al. (1985) Observation of toroidal coupling for low-n Alfven modes in the TCA tokamak, *Phys. Rev. Lett.* **54**, 1671.

4. Collins G.A. et al. (1986) The Alfven wave spectrum as measured on a tokamak, *Phys. Fluids* **29**, 2260.

5. Cross R.C., Blackwell B.D., Brennan M.H., Borg G.G. and Lehane J.A. (1982) Magnetic probe measurements of torsional Alfven waves in the TORTUS tokamak, *Proc. 3rd Int. Symp. on Heating in Toroidal Plasmas*, Grenoble, Vol. 1, p173.

6. DeChambrier A. et al. (1982) Alfven wave coupling experiments on the TCA tokamak, *Plasma Phys.* **24**, 893.

7. DeChambrier A. et al (1983) Measurements of electron and ion heating by Alfven waves in the TCA tokamak, *Plasma Phys.* **25**, 1021.

8. Evans T.E. et al. (1984) Direct observation of the structure of global Alfven eigenmodes in a tokamak plasma, *Phys. Rev. Lett.* **53**, 1743.

9. Knox S.O., Paoloni F.J. and Kristiansen M. (1975) Helical antenna for exciting azimuthally asymmetric Alfven waves, *J. Appl. Phys.* **46**, 2516.

10. Mutoh T. et al. (1979) Investigation of electron heating due to the kinetic shear Alfven wave in the toroidal plasma, *J. Phys. Soc. Japan* **47**, 642.

11. Obiki T. (1977) Alfven-wave heating experiment in the Heliotron-D, *Phys. Rev. Lett.* **39**, 812.

12. Shvets O.M. et al. (1986) Absorption of Alfven waves and plasma production in the Omega and Uragan-3 toroidal devices, *Nucl. Fusion* **26**, 23.

13. Tsushima A., Amagishi Y. and Inutake M. (1982) Observation of spatial Alfven resonance, *Phys. Lett.* **88A**, 457.

14. Witherspoon F.D., Prager S.C. and Sprott J.C. (1984) Experimental observation of the shear Alfven resonance in a tokamak, *Phys. Rev. Lett.* **53**, 1559.

G. Magnetically guided Alfven waves

1. Borg G.G. et al. (1985) Guided propagation of Alfven waves in a toroidal plasma, *Plasma Phys. Contr. Fusion* **27**, 1125.

2. Borg G.G. and Cross R.C. (1987) Guided propagation of Alfven and ion-ion hybrid waves in a plasma with two ion species, *Plasma Phys. Contr. Fusion* **29**, 681.

3. Cross R.C. (1983) Experimental observations of localised Alfven and ion acoustic waves in a plasma, *Plasma Phys.* **25**, 1377.

4. Cross R.C. (1985) Demonstration of wave propagation in a periodic structure, *Am. J. Phys.* **53**, 563.

5. Fejer J.A. and Lee K.F. (1967) Guided propagation of Alfven waves in the magnetosphere, *J. Plasma Phys.* **1**, 387.

6. Fejer J.A. (1981) Guided magnetospheric propagation of Alfven waves in the Pc4-5 frequency range, *J. Geophys. Res.* **86**, 5614.

7. Heyvaerts J. and Priest E.R. (1983) Coronal heating by phase-mixed shear Alfven waves, *Astron. Astrophys.* **117**, 220.

8. Musielak Z.E. (1986) Relationship between directions of wave and energy propagation for cold plasma waves, *J. Plasma Phys.* **36**, 341.

9. Shoucri M.M., Morales G.J. and Maggs J.E. (1985) Generation of Alfven waves caused by thermal modulation of the electrical conductivity, *Phys. Fluids* **28**, 2458.

10. Sy W.N.C. (1984) Influence of resistivity on localised shear Alfven wave propagation in a nonuniform plasma, *Plasma Phys. Contr. Fusion* **26**, 915.

11. Walker A.D.M. (1977) The phase velocity, ray velocity and group velocity surfaces for a magneto-ionic medium, *J. Plasma Phys.* **17**, 467.

12. Weinberg S. (1962) Eikonel method in magnetohydrodynamics, *Phys. Rev.* **126**, 1899.

H. Alfven wave spectrum

1. Appert K., Gruber R. and Vaclavik J. (1974) Continuous spectra of a cylindrical magnetohydrodynamic equilibrium, *Phys. Fluids* **17**, 1471.

2. Appert K., Vaclavik J. and Villard L. (1984) Spectrum of low- frequency, non-axisymmetric oscillations in a cold, current-carrying plasma column, *Phys. Fluids* **27**, 432.

3. Chance M.S., Green J.M., Grimm R.C. and Johnson J.L. (1977) Study of the MHD spectrum of an elliptic plasma column, *Nucl. Fusion* **17**, 65.

4. Cheng C.Z., Chen L. and Chance M.S. (1985) High-n and resistive shear Alfven waves in tokamaks, *Annals of Physics* **161**, 21.

5. Cheng C.Z. and Chance M.S. (1986) Low-n shear Alfven spectra in axisymmetric toroidal plasmas, *Phys. Fluids* **29**, 3695.

6. Davis B. (1984) The spectrum of resistive MHD equations, *Phys. Lett.* **100A**, 144.

7. Dewar R.L. and Davies B. (1984) Bifurcation of the resistive Alfven wave spectrum, *J. Plasma Phys.* **32**, 443.

8. D'Ippolito D.A. and Goedbloed J.P. (1980) Mode coupling in a toroidal sharp-boundary plasma - I. Weak-coupling limit, *Plasma Phys.* **22**, 1091. II. Strong-coupling limit, *Plasma Phys.* **25**, 537.

9. Goedbloed J.P. (1975) Spectrum of ideal magnetohydrodynamics of axially symmetric toroidal systems, *Phys. Fluids* **18**, 1258.

10. Kerner W., Lerbinger K., Gruber R. and Tsunematsu T. (1985) Normal mode analysis for resistive cylindrical plasmas, *Computer Phys. Communications* **36**, 225.

11. Kieras C.E. and Tataronis J.A. (1982) The shear Alfven continuous spectrum of axisymmetric toroidal equilibria in the large aspect ratio limit, *J. Plasma Phys.* **28**, 395.

12. Lortz D. and Spies G. (1984) Spectrum of a resistive plasma slab, *Phys. Lett.* **101A**, 335.

13. Mahajan S.M., Ross D.W.and Chen G.L. (1983) Discrete Alfven eigenmode spectrum in magnetohydrodynamics, *Phys. Fluids* **26**, 2195.

14. Riedel K.S. (1986) The spectrum of resistive viscous magnetohydrodynamics, *Phys. Fluids* **29**, 1093 and 2986.

15. Ryu C.M. and Grimm R.C. (1984) The spectrum of resistive MHD modes in cylindrical plasmas, *J. Plasma Phys.* **32**, 207.

I. ICRF Heating (theory)

1. Cattanei G. and Croci R. (1977) Fast-wave heating of a toroidal plasma near the ion cyclotron frequency, *Nucl. Fusion* **17**, 239.

2. Colestock P.L. and Kashuba R.J. (1983) The theory of mode conversion and wave damping near the ion cyclotron frequency, *Nucl. Fusion* **23**, 763.

3. Hellston T. and Tennfors E. (1984) Resonance surfaces in the ion cyclotron frequency range in toroidal systems, *Physica Scripta* **30**, 341.

4. Messiaen A.M., Vandenplas P.E., Weynants R.R. and Koch R. (1975) Bounded-plasma theory of radio-frequency heating of toroidal machines, *Nucl. Fusion* **15**, 75.

5. Perkins F.W. (1977) Heating tokamaks via the ion cyclotron and ion-ion hybrid resonances, *Nucl. Fusion* **17**, 1197.

6. Porkolab M. (1981) Review. Radio frequency heating of magnetically confined plasma, in *Fusion* (Ed. Teller E.) Academic Press, New York.

7. Scharer J.E., McVey B.D. and Mau T.K. (1977) Fast wave ion-cyclotron and first-harmonic heating of large tokamaks, *Nucl. Fusion* **17**, 297.

8. Stix T.H. (1975) Fast-wave heating of a two-component plasma, *Nucl. Fusion* **15**, 737.

9. Stix T.H. and Swanson D.G. (1983) Propagation and mode-conversion for waves in nonuniform plasmas, in *Handbook of Plasma Physics* , edited by Galeev A. and Sudan R.N. (North-Holland) Vol. 1, p335.

10. Swanson D.G. and Ngan Y.C. (1975) Warm-plasma effects on fast-Alfven-wave cavity resonances, *Phys. Rev. Lett.* **35**, 517.

11. Swanson D.G. (1976) Mode conversion and tunneling at the two-ion hydrid resonance, *Phys. Rev. Lett.* **36**, 316.

12. Swanson D.G. (1985) Review. Radio frequency heating in the ion-cyclotron range of frequencies, *Phys. Fluids* **28**, 2645.

13. Takahashi H. (1977) ICRF heating in tokamaks, *Journal de Physique* **38**, 171 (Princeton report PPPL-1374).

14. Weynants R.R. (1974) Ion heating at twice the ion-cyclotron frequency in reactor-oriented machines, *Phys. Rev. Lett.* **33**, 78.

J. ICRF Heating experiments

The following list is a small sample only. A large amount of work in this field is reported in conference proceedings (Sect. **A**). A good introduction to the subject is the review by Adam (J2).

1. Adam J. (1984) ICRF heating in TFR and the problem of impurity release, *Plasma Phys. Contr. Fusion* **26**, 165.

2. Adam J. (1987) Review of tokamak plasma heating by wave damping in the ion cyclotron range of frequency, *Plasma Phys. Contr. Fusion* **29**, 443.

3. Akiyama H., Wong K.L., Gahl J., Kristiansen M. and Hagler M. (1987) Fast Alfven wave propagation in a deuterium-hydrogen tokamak plasma, *Plasma Phys. Contr. Fusion* **29**, 93.

4. Hosea J. et al. (1982) High power ICRF heating on PLT and extrapolation to future devices, *Proc. 3rd Int. Symp. on heating in toroidal plasmas,* Grenoble, Vol. 1, p213.

5. Ida K., Naito M., Shinohara S. and Miyamoto K. (1983) Direct measurement of fast magnetosonic wave near the ion-ion hybrid resonance layer by magnetic probes, *Nucl. Fusion* **23**, 1259.

6. Lapierre Y. (1983) Magnetosonic wave propagation in the mode conversion regime, *J. Plasma Phys.* **29**, 223.

7. Messian A.M., Weynants R.R., Bhatnagar V.P. and Vandenplas P.E. (1979) Ion-ion hybrid damping of magneto-acoustic resonances, *Phys. Lett.* **71A**, 431.

8. Messian A.M. et al. (1986) Ion cyclotron resonance heating in TEXTOR, *Plasma Phys. Contr. Fusion* **28**, 71.

9. Ono M., Wong K.L. and Wurden G.A. (1983) Harmonic launching of ion Bernstein waves via mode transformation, *Phys. Fluids* **26**, 298.

10. Ono M., Wurden G.A. and Wong K.L. (1984) Efficient ion heating via finite-Larmor-radius ion-cyclotron waves in a plasma, *Phys. Rev. Lett.* **52**, 37.

11. Ono M. et al. (1985) Ion-Bernstein-wave heating in the JIPPT-II-U tokamak plasma, *Phys. Rev. Lett.* **54**, 2339.

12. Shinohara S., Naito O. and Miyamoto K. (1986) Dependence of resistive loading on antenna orientation for ICRF waves in a tokamak, *Nucl. Fusion* **26**, 1097.

13. Takahashi H. et al. (1977) Ion heating in ATC tokamak in the ion-cyclotron range of frequencies, *Phys. Rev. Lett.* **39**, 31.

K. Magnetoacoustic waves

1. Blackwell B.D. and Cross R.C. (1979) Measurements of the impedance of a magneto-acoustic wave-launching antenna, *J. Plasma Phys.* **22**, 499.

2. Brennan M.H., McCarthy A.L. and Sawley M.L. (1980) Determination of equilibrium plasma density and magnetic field profiles using magneto-acoustic oscillations, *Plasma Phys.* **22**, 77.

3. Cross R.C. (1970) Radio-frequency heating of a plasma, *Canadian J. Phys.* **48**, 2888.

4. Lindberg L. and Danielsson L. (1963) Magneto-acoustic oscillations in a cylindrical plasma column, *Phys. Fluids* **6**, 736.

5. Ritz Ch.P., Hoegger B.A., Sayasov Y.S. and Schneider H. (1982) On the turbulence in a magnetized plasma and its influence on the magneto- acoustic resonance, *Helv. Physica Acta* **55**, 354.

L. Experimental techniques

1. Anderson J.M. (1971) Wide frequency range current transformers, *Rev. Sc. Instr.* **42**, 915.
2. Beiersdorfer P. and Clothiaux E.J. (1983) High-frequency magnetic measurements using small inductive probes, *Am. J. Phys.* **51**, 1031.
3. Borg G. (1987) Broadband amplitude independent phase measuring system, *J. Phys. E: Sc. Instr.* **20**, 1216.
4. Cross R.C. (1986) Current transformers, *Am. J. Phys.* **54**, 1110.
5. Cross R.C. and Collins G.A. (1981) Compensated RC integrators, *Am. J. Phys.* **49**, 479.
6. Hayward W.H. (1982) *Introduction to radio frequency design* , Prentice-Hall, N.J.
7. Ruthroff C.L. (1959) Some broad-band transformers, *Proc. IRE* **47**, 1337.
8. Zverev A.I. (1967) *Handbook of filter synthesis* , Wiley, N.Y.

M. Space plasmas

1. Ansari I.A. and Fraser B.J. (1986) A multistation study of low latitude Pc3 geomagnetic pulsations, *Planet. Space Sci.* **34**, 519.
2. Cesarsky C.J. (1980) Cosmic-ray confinement in the Galaxy, *Ann. Rev. Astr. Astophys.* **18**, 289.
3. Davila J.M. (1987) Heating of the solar corona by the resonant absorption of Alfven waves, *Astron. Astrophys.* **317**, 514.
4. Fraser B.J. (1985 Review) Observations of ion cyclotron waves near synchronous orbit and on the ground, *Space Science Reviews* **42**, 357.
5. Hollenbach D.J. and Thronson H.A. (Eds. 1986) *Interstellar Processes*, D. Reidel Publishing Co., Dordrecht, Holland.
6. Jacobs J.A. (1970) *Geomagnetic Micropulsations*, Springer-Verlag, Berlin.
7. Kivelson M.G. and Southwood D.J. (1986) Coupling of global magnetospheric MHD eigenmodes to field line resonances, *J. Geophys. Res.* **91**, 4345.
8. Kuperus M., Ionson J.A. and Spicer D.S. (1981) On the theory of coronal heating mechanisms, *Ann. Rev. Astron. Astrophys.* **19**, 7.

9. Lanzerotti L.J. and Southwood D.J. (1979) Hydromagnetic Waves, in *Solar System Plasma Physics*, Vol III, p.109, Ed. by Lanzerotti L.T., Kennel C.F. and Parker E.N., North-Holland.

10. Mein N. and Mein P. (1980) Mechanical flux in the solar chromosphere, *Astron. Astrophys.* **84**, 96-105.

11. Musielak Z.E. and Rosner R. (1987) On the generation of magnetohydrodynamic waves in a stratified and magnetised fluid. I. Vertical propagation, *Astrophys. Jnl.* **315**, 371.

12. Nocera L., Leroy B. and Priest E.R. (1984) Phase mixing of propagating Alfven waves, *Astron. Astrophys.* **133**, 387.

13. Parker E.N. (1979) *Cosmical magnetic fields*, Oxford Univ. Press Oxford.

14. Priest E.R. (1982) *Solar magnetohydrodynamics*, Riedel, Dordrecht, Holland.

15. Priest E.R. (Ed. 1985) *Solar system magnetic fields*, Riedel, Dordrecht, Holland.

16. Schmerling E.R., Cowley S.W. and Reiff P.H. (Eds., 1985) *Magnetospheric and ionospheric plasmas*, Pergamon, Oxford.

17. Southwood D.J. (1974) Some features of field line resonances in the magnetosphere, *Planet. Space Sci.* **22**, 483.

18. Southwood D.J. and Hughes W.J. (1983 Review) Theory of hydromagnetic waves in the magnetosphere, *Space Science Reviews* **35**, 301.

19. Takahashi K. and McPherron R.L. (1984) Standing hydromagnetic oscillations in the magnetosphere, *Planet. Space Sci.* **32**, 1343.

20. Wentzel D.G. (1974) Cosmic-ray propagation in the Galaxy: Collective effects, *Ann. Rev. Astron. Astrophys.* **12**, 71.

Index